Lecture Notes in Computer Science 13488

More information about this series at https://link.springer.com/bookseries/558

Andreas Bollin · Gerald Futschek (Eds.)

Informatics in Schools

A Step Beyond Digital Education

15th International Conference on Informatics in Schools:
Situation, Evolution, and Perspectives, ISSEP 2022
Vienna, Austria, September 26–28, 2022
Proceedings

 Springer

Editors
Andreas Bollin 🆔
Universität Klagenfurt
Klagenfurt, Kärnten, Austria

Gerald Futschek 🆔
TU Wien
Vienna, Austria

ISSN 0302-9743 ISSN 1611-3349 (electronic)
Lecture Notes in Computer Science
ISBN 978-3-031-15850-6 ISBN 978-3-031-15851-3 (eBook)
https://doi.org/10.1007/978-3-031-15851-3

This Springer imprint is published by the registered company Springer Nature Switzerland AG
The registered company address is: Gewerbestrasse 11, 6330 Cham, Switzerland

Preface

This volume contains all the research, best practice, and country and experience reports presented at the 15th International Conference on Informatics in Schools: Situation, Evolution, and Perspectives (ISSEP 2022). The conference was held at TU Wien, Austria, during September 26–28, 2022, in cooperation with the University of Klagenfurt, Austria. Invitees included not only researchers in the field of computer science didactics but also computer scientists, teachers, stakeholders from industry, and staff from the Federal Ministry of Education, Science, and Research.

The conference series started in Klagenfurt, Austria, in 2005, when information and communication technologies were increasingly making their way into the classroom and beginning to displace traditional computer science teaching. In order to educate decision-makers, Roland Mittermeir initiated ISSEP, planned initially as a one-time international event. However, it did not stop there, and the ISSEP conference has so far taken place in Vilnius, Lithuania (2006), Torun, Poland (2008), Zurich, Switzerland (2010), Bratislava, Slovakia (2011), Oldenburg, Germany (2013), Istanbul, Republic of Türkiye (2014), Ljubljana, Slovenia (2015), Münster, Germany (2016), Helsinki, Finland (2017), St. Petersburg, Russia (2018), Larnaca, Cyprus (2019), Tallinn, Estonia (2020), and Nijmegen, The Netherlands (2021).

In the meantime, something very interesting is happening again in our school systems: subjects like "digital literacy" or "media literacy" are making their way in, complementing or partially replacing computer science education. The current ISSEP conference reacted to this trend and therefore invited computer scientists, media didactics, and representatives of politics and industry to a discussion round on the topic "Media Education or Computer Science? Quo Vadis, School Teaching?".

The conference makes an equally strong effort to promote young researchers, offering a Doctoral Consortium the day before the conference. In total, 11 Ph.D. students presented and discussed their research on September 25, 2022. They received assistance from international peers and introduced new ideas to their research careers.

The conference received a total of 57 submissions. Of these, 25 submissions were full papers, four short papers, eight workshop proposals, nine poster proposals, and 11 Doctoral Consortium topics. Each submission was reviewed in a double-blind review process and was evaluated, discussed, and selected by at least three reviewers together with the program chairs, except for the workshop proposals and the Doctoral Consortium where two to three reviewers reviewed and selected the topics. The reviewers selected 12 submissions for publication in the LNCS proceedings, resulting in an acceptance rate (for full research papers) of 48%. The decision process was performed electronically using the EasyChair conference management system.

Past ISSEP conferences attracted submissions on various computer science didactics/school teaching content. This year, too, there were contributions in many areas. However, the topics dealing with computational thinking, primary education, and Bebras tasks slightly outweighed the others. There were also contributions dealing with curricula and examples of school practice.

Finally, we would like to thank everyone who made this conference possible: the authors with their submissions, the many members of the Program Committee who did a fantastic job, the sponsors, all the participants of the conference, and the local organization team.

September 2022

Andreas Bollin
Gerald Futschek

Organization

Conference Chairs

Andreas Bollin	University of Klagenfurt, Austria
Gerald Futschek	TU Wien, Austria

Steering Committee

Andreas Bollin (Chair)	University of Klagenfurt, Austria
Valentina Dagienė	Vilnius University, Lithuania
Yasemin Gülbahar	Ankara University, Republic of Türkiye
Juraj Hromkovič	ETH Zurich, Switzerland
Ivan Kalas	Comenius University, Slovakia
Erik Barendsen	Radboud University and Open University, The Netherlands
Sergei Pozdniakov	Saint Petersburg Electrotechnical University, Russia

Program Committee

Andreas Bollin (Chair)	University of Klagenfurt, Austria
Peter Antonitsch	University of Klagenfurt, Austria
Andrej Brodnik	University of Ljubljana, Slovenia
Špela Cerar	University of Ljubljana, Slovenia
Christian Datzko	Wirtschaftsgymnasium und Wirtschafts-Mittelschule Basel, Switzerland
Monica Divitini	Norwegian University of Science and Technology, Norway
Gerald Futschek	TU Wien, Austria
Juraj Hromkovič	ETH Zurich, Switzerland
Mile Jovanov	Ss. Cyril and Methodius University of Skopje, North Macedonia
Kaido Kikkas	Tallinn University, Estonia
Dong Yoon Kim	Ajou University, South Korea
Dennis Komm	ETH Zurich, Switzerland
Mart Laanpere	Tallinn University, Estonia
Martina Landman	TU Wien, Austria
Peter Larsson	University of Turku, Finland
Marina Lepp	University of Tartu, Estonia

Nina Lobnig	University of Klagenfurt, Austria
Birgy Lorenz	Tallinn University, Estonia
Piret Luik	University of Tartu, Estonia
Maia Lust	Tallinn University, Estonia
Kati Mäkitalo	University of Oulu, Finland
Tilman Michaeli	TU Munich, Germany
Mattia Monga	Università degli Studi di Milano, Italy
Tauno Palts	University of Tartu, Estonia
Stefan Pasterk	University of Klagenfurt, Austria
Hans Põldoja	Tallinn University, Estonia
Sergei Pozdniakov	Saint Petersburg Electrotechnical University, Russia
John-Paul Pretti	University of Waterloo, Canada
Ralf Romeike	Freie Universität Berlin, Germany
Barbara Sabitzer	Johannes Kepler Universität Linz, Austria
Carsten Schulte	University of Paderborn, Germany
Giovanni Serafini	ETH Zurich, Switzerland
Vipul Shah	ACM India CSpathshala Education Initiative, India
Gabrielė Stupurienė	Vilnius University, Lithuania
Reelika Suviste	University of Tartu, Estonia
Maciej Syslo	Nicolaus Copernicus University in Toruń, Poland
Michael Weigend	University of Münster, Germany
Albin Weiss	University of Klagenfurt, Austria
Markus Wieser	University of Klagenfurt, Austria

Doctoral Consortium Committee

Valentina Dagienė (Chair)	Vilnius University, Lithuania
Andreas Bollin	University of Klagenfurt, Austria
Gerald Futschek	TU Wien, Austria
Barbara Sabitzer	Johannes Kepler Universität Linz, Austria
Carsten Schulte	University of Paderborn, Germany

Local Organizers

Gerald Futschek (Chair)	TU Wien, Austria
Franziska Tiefenthaller (Organization)	TU Wien, Austria
Martin Krajiczek (IT Support)	TU Wien, Austria

Peter Kompatscher   TU Wien, Austria
   (Event Management)
Stefan Pasterk (Publicity Chair)   University of Klagenfurt, Austria
Melanie Ottowitz   University of Klagenfurt, Austria
   (Publicity Support)

Contents

State of Research

Informatics Education in German Primary School Curricula

Christin Nenner[(✉)] and Nadine Bergner

TU Dresden, Dresden, Germany
{christin.nenner,nadine.bergner}@tu-dresden.de

Abstract. The world is permeated by informatics, even the youngest children encounter informatics systems and the respective phenomena. Consequently, even primary school children must acquire the necessary informatics competencies to act in this world in a self-determined manner. A study on the integration of informatics education in European education systems by the Committee on European Computing Education showed that in 2017, none of the German states provided informatics competencies in primary school. In the following years, as the study Reviewing Computational Thinking in Compulsory Education (2022) shows, progress towards more informatics education in primary schools has been made in various European countries. For example, in Sweden, informatics content has been integrated into the curricula. However, the current situation in Germany was not analyzed in this study.

The aim of this paper is to analyze the current state of informatics education in Germany's primary schools, comparing it with pioneering countries, and to identify potential for development. For this purpose, 53 curricula of different subjects in all 16 German federal states (as there is no specific subject for informatics) were systematically scanned for informatics content. Subsequently, the informatics competencies found were categorized with respect to the CSTA K-12 standards in order to work out which subject areas are currently dealt with to what extent.

The results show eight out of 16 German federal states integrated informatics competencies as a mandatory part of primary level education. With regard to the main concepts, Algorithms and Programming and Computing Systems are particularly addressed, whereas the area of Networks and the Internet is only assigned in one federal state. The greatest potential for development is to be seen in teacher training so that the competencies required in the curricula can be developed professionally with the children.

Keywords: Informatics education · Primary school · Primary school curricula · German informatics education · CSTA K-12 CS standards

"The project underlying this article is part of the "Qualitätsoffensive Lehrerbildung", a joint initiative of the Federal Government and the Länder which aims to improve the quality of teacher training. The programme is funded by the Federal Ministry of Education and Research. The authors are responsible for the content of this publication".

A. Bollin and G. Futschek (Eds.): ISSEP 2022, LNCS 13488, pp. 3–14, 2022.
https://doi.org/10.1007/978-3-031-15851-3_1

1   Introduction

Already in 2017, the Committee on European Computing Education (CECE) demanded that all students should receive continuous informatics education, preferably starting in primary school [22, p. 3]. The European Digital Education Action Plan 2021–2027 also advocates "a focus on inclusive high-quality computing education [informatics] at all levels of education" [7, p. 15], thus including primary school. As part of the study "Informatics Education in Europe: Are We All In The Same Boat?" [22] published in 2017, the situation of informatics education in primary schools in Germany was also examined. At that time, according to the study, none of the curricula of the 16 German states taught informatics content in primary school. Since then, in addition to the aforementioned European action plan in Germany, several other impulses for integrating informatics content into (primary) school curricula have been published. At the end of 2021, a more informatics-oriented supplement to the strategy "Education in the Digital World" from the Standing Conference of the Ministers of Education and Cultural Affairs (KMK) was published [14]. At the beginning of 2022, the German Rectors' Conference further advocated informatics education in all teacher training programs [9].

According to the study "Reviewing Computational Thinking in Compulsory Education" by the European Commission's Joint Research Centre, 17 European countries (for example, Austria, Greece, and Finland) have "introduced basic computer science [informatics] concepts as a compulsory subject in both primary and lower secondary education" [8, p. 5]. Germany was not among the European countries studied in this regard. The situation of teaching informatics in secondary schools is already analyzed and made visible in the Informatics Monitor [20]. The situation in primary school is not addressed there.

Therefore, the aim of this paper is to analyze the current state of informatics education in primary schools in all German federal states, to compare it internationally and to identify development potentials.

1.1   International State of Informatics Education in Primary Schools

The 2017 published CECE report analyzed basic informatics education in schools, digital literacy orientation, and related teacher training across Europe [22]. For this purpose, the CECE defines informatics as "a distinct scientific discipline, characterised by its own concepts, methods, body of knowledge, and open issues. It covers the foundations of computational structures, processes, artefacts and systems; and their software designs, their applications, and their impact on society" [22, p. 3]. This report presents "a mapping across 55 administrative units (countries, nations, and regions) of Europe (including Israel) with autonomous educational systems" [22, p. 3]. The data shows that in six out of 55 countries/regions (Croatia, Slovenia, Ukraine, and all countries of the United Kingdom (UK)) students' first contact with informatics is in primary school [22, p. 12]. Outside the EU, informatics content has been a topic in schools for many years, for example in Macedonia [12] and Turkey [10].

Since the publication of the report [22] in 2017, a lot has happened in some countries regarding the teaching of informatics:

- In Sweden, curricula (grades 1 to 9) for math, technology, and social studies that include informatics content have been mandatory since 2018. Here, the informatics content is integrated into existing subjects. "The revision puts increased attention on digital technology and the need for developing an understanding for how computers and networks work." [11, p.121]. Clear algorithmic instructions as the basis of programming are primarily covered in mathematics classes [11, p.122]. In technology, the structure and functioning of informatics systems and the control of technical devices through programming are taken up [11, p. 122]. In social sciences, the Impact of Informatics is addressed [11, p. 122].
- In Lithuania, a draft curriculum for informatics for grades 1–4 of primary school was developed in 2016. This covers, among other things, Data and Information and Algorithms and Programming [5, p. 89].
- In 2019, Denmark launched a three-year field trial of informatics teaching in primary schools [23, p. 2].
- At the same time, Poland also introduced an informatics curriculum in primary school [23, p. 2].

1.2 CSTA K-12 Computer Science Standards

The CSTA K-12 Computer Science (CS) Standards were published in 2017 as concepts and matching learning objectives that map the foundations for a complete informatics curriculum from grades 1 through 12. The authors most important goal is to introduce fundamental concepts of informatics to all students beginning in primary school. These standards are divided into five basic concepts: 1. Computing Systems, 2. Networks and the Internet, 3. Data and Analysis, 4. Algorithms and Programming and 5. Impacts of Computing [4].

2 Description of Research Object

In order to classify the research presented below internationally, the German school system will be outlined and the range of subjects taught in primary schools will be explained.[1]

2.1 German Education System

In Germany, educational sovereignty lies within the 16 federal states[2] (Baden-Württemberg (BW), Bavaria (BY), Berlin (BE) + Brandenburg (BB), Bremen (HB), Hamburg (HH), Hesse (HE), Mecklenburg-Western Pomerania (MV),

---

[1] If German terms are used, these are marked in italics.
[2] Detailed information can be found within the document: https://www.kmk.org/fileadmin/Dateien/pdf/Eurydice/Bildungswesen-engl-pdfs/dossier_en_ebook.pdf.

Lower Saxony (NI), North Rhine-Westphalia (NW), Rhineland-Palatinate (RP), Saarland (SL), Saxony (SN), Saxony-Anhalt (ST), Schleswig-Holstein (SH), Thuringia (TH)). They all have independent educational authorities and independent curricula that differ in content as well as structure. Even the canon of subjects itself varies between the states. In 14 states, primary school covers grades 1 to 4. Only in BE and BB primary school continues up to grade 6. These states have a common framework curriculum and are therefore considered as one in the following. In some states, there are also overarching documents that apply to all primary school curricula. In Germany no independent school subject for informatics content exists in primary school in any of the federal states. When informatics content is addressed, it is integrated into other primary school subjects.

2.2 Overview of Primary School Curricula

The curricula of the three core subjects German, math and science (*Sachunterricht*) were selected for investigation in the context of this paper. In the states of BY (*Heimat- und Sachunterricht*) and TH (*Heimat- und Sachkunde*), the naming of the subject science differs. In some federal states, there is an additional subject in which students can develop craft and technical skills: *Kunst/Werken* (BW), *Werken und Gestalten* (BY), *Werken* (MZ, SN, TH), *Gestaltendes Werken* (NI), *Gestalten* (ST), and *Technik* (SH). This selection will be referred to as technology in the following. Due to the frequent technical orientation of this subject and because an extra learning area for informatics content was created in the subject of *Werken* in Saxony in 2019, the curricula for technology were also examined in addition to the curricula for the three core subjects.

A total of 53 curricula were considered for the document analysis. For all of them, the date of publication or entry into force of this version of the curricula and the presence of the respective subjects were included, depending on the information provided (see Table 1).

2.3 Research Questions and Design

A document analysis of the current curricula of all federal states was conducted to investigate the following research questions:

- In which federal states is informatics education included as an overarching goal in current primary school curricula?
- In which federal states is informatics content mentioned within learning or subject areas?
- What informatics competencies (or content) are targeted?
- To which concepts of the CSTA K-12 CS Standards can the informatics content be assigned?

The curricula are available as PDF files and were scanned for specific keywords using the search function of a PDF reader.

Table 1. Presentation of the subjects taught in the 16 federal states with the year of publication or entry into force. The presence of informatics content including the assignment to the respective subject is highlighted by bold and gray marking. In the case of technology, the subject name is included.

	German	Math	Science	Technology
BW	2016	2016	2016	2016 | *Kunst/Werken*
BY	2014	2014	2014 | *Heimat- und Sachunterricht*	—
BE + BB	2015	2015	**2015**	—
HB	2004	2004	2007	—
HH	2011 + 2020	2011 + **2020**	**2011**	—
HE	1995	1995	1995	—
MV	2020	**Not specified**	2020	Not specified | *Werken*
NI	2017	2017	2017	2006 | *Gestaltendes Werken*
NW	2021	**2021**	**2021**	—
RP	2005	2014	2006	—
SL	2009	2009	2010	—
SN	2019	2019	2019	**2019** | ***Werken***
ST	2019	**2019**	2019	2019 | *Gestalten*
SH	2018	2018	2019	**2021** | ***Technik***
TH	2010	2010	2015 | *Heimat- und Sachkunde*	2010 | *Werken*

To examine the curricula, they were screened for 16 keywords. The keywords were aligned with both the concepts of the CSTA K-12 CS Standards [4] and the recommendations on competencies for the primary level from the German Informatics Society (content areas and keywords of competency expectations) [3]. The goal of searching for additional keywords (besides *Informatik, informatisch*) is to locate informatics content even if it is not labeled as such. All text passages found with the help of the keywords were manually examined to determine whether they contained informatics content. Since other studies sometimes do not mention which content is mandatory or optional, only mandatory learning areas are considered here. The following is a list of the English equivalents of the German keywords used for the analysis with German keywords in [ ].

- algorithm, algorithmic [*Algorithmen, algorithmisch*]
- automaton, incl. automation [*Automat, Automatisierung*]
- binary [*binär*]
- coding [*codieren, Codierung*]
- computer [*Computer*]
- data [*Daten*]
- digital [*digital*]
- encryption [*verschlüsseln, Verschlüsselung*]
- functionality [*Funktionsweise*]
- informatic(s)/computing, incl. informatics/computing system [*Informatik, informatisch, Informatiksystem*]
- input, IPO model (Input-Process-Output) [*Eingabe, EVA-Prinzip*]
- internet [*Internet*]

- network [*Netzwerk*]
- processing [*Verarbeitung*]
- program, programming, incl. programming language [*Programm, program-mieren, Programmiersprache*]
- robot [*Roboter*]

3   Results

In eight out of 16 German states, informatics content is integrated in mandatory learning areas in primary school curricula (see Table 2). In four states (BE, BB, HH, NW), it is integrated in science, in four states (HH, MV, NW, ST) in math, and in two states (SN, SH) in technology. In six of these eight states, informatics content is integrated in exactly one subject. Only two states (HH, NW) have integrated informatics content in two subjects (science and math). There is no federal state in which there is an independent subject for informatics content. Federal states in which at least one curriculum contains informatics content have for the most part (six out of eight) updated or published their curricula in 2018 or later. From a publication date of after 2017/2018, many curricula refer to the strategy "Education in the Digital World" from KMK [13]. In some of the curricula, this document is relied upon for the introduction of explicit informatics content. However, it cannot be assumed that a curriculum revised after 2017 necessarily contains informatics content (e.g. NI - German, math, science), nor that a publication date before 2017 means that no informatics content can be integrated (e.g. BE, BB - science | HH - science).

In the context of the evaluation, it is considered separately whether the keywords *Informatik, informatisch* are referred to in other sections (e.g. the preliminary remarks, the overarching educational objectives or even the didactic principles) in addition to the mention in informatics content. Table 2 shows that in five out of 16 federal states the mentioned terms appear in other sections of the curricula in addition to the learning areas. Here, the main references are to informatics education, STEM[3] education, and connectivity to informatics. In two of the five federal states (NI, RP), STEM or informatics education is mentioned in a superordinate way, but there is no explicit informatics content.

In the next step, it is shown which specific informatics content is integrated in the curricula of the federal states. For this purpose, all text sections found by means of keyword search were analyzed. All text sections that did not refer to informatics content were discarded (e.g., when searching with the term digital: "read all times on analog and digital clocks"). For each state, all informatics content found was collected. The respective collection was then mapped to the CSTA K-12 CS Standards [4]. In Table 2 below the summarized informatics content per state is presented.

The distribution of CSTA K-12 CS standards across states with informatics content shows that the informatics content found can be assigned to three, four, or five concepts for half of the eight states (HH, NW, SN, SH). All four

---

[3] Science, technology, engineering, and math.

Table 2. Overview of informatics mentioned in overarching parts and the summarized informatics content in mandatory learning areas to be addressed per federal state, assigned to the subjects and including the naming of the learning areas. The assignment to the concepts of the CSTA K-12 CS Standards [4] is presented.

Federal State	Informatics mentioned in overarching parts	Informatics content in mandatory learning areas to be addressed	Computing Systems	Networks and the Internet	Data and Analysis	Algorithms and Programming	Impacts of Computing
BE + BW	—	**Science** (Topic: Child, Theme: What do we know about? | grades 1-4 [15, p. 31]): – Computers and the internet: How do computers work?	× ×				
HH	—	**Math** (Topic: Solving problems mathematically | grades 1-4 [2, p. 5])): – Computational thinking (logical series, codes, binary code, structure of algorithms, regular sequences) – Implications of automation for one's own reality of life – Formalizing and describing problems – Basic skills in programming **Science** (Topic: Technology - Orientation in our world | grades 3-4 [1, p. 24 f]): – Components of computers and automata – Input-process-output – Media for data storage incl. data sets – Internet as a form of networking computers and as a technology through which texts, images and sounds are transmitted	×	×	×	×	×
MV	—	**Science** (Topic: Using, evaluating, and producing media | grades 3-4 [16, p. 35]): – Following and formulating algorithms				×	
NI	**Math, science:** contribution to interdisciplinary educational areas → STEM education	—					

(*continued*)

Table 2. (*continued*)

NW	**Science:** tasks and goals of science → informatics education	**Math** (Topic: Numbers and operations | grades 3-4 [18, p. 86]): – Conversion of numbers between decimal and binary system **Science** (Topic: Democracy and society | grades 3-4 [18, p. 185]): – Differentiation between coding and encoding of data – Possibilities for protecting personal data **Science** (Topic: Engineering, digital technology, and work | grades 3-4 [18, p. 192]): – Input-process-output as a basic principle of data processing in informatics systems – Programming a sequence	×		×	×	×
RP	**Math:** preliminary remarks → informatics education, STEM education	—					
SN	**All curricula:** overarching educational goal → informatics (pre)education	***Werken*** (Topic: Encountering robots and automata | grade 4 [21, p. 14]): – Application of robots and automata → Input-process-output – Application of a simplified software development cycle including the programming of a simple procedure – Transfer of knowledge to the implementation of a concrete task (tracing input-process-output, building a model, evaluating the implementation)	×			×	×
ST	—	**Math** (Topic: Numbers and operations | grades 3-4 [19, p. 7]): – Tasks in factual situations (e.g., for encrypting data or securing access)					×
SH	**Science:** didactic guidelines of science → connectivity to informatics	***Technik*** (Topic: Information and communication | grades 1-4 [17, p. 16]): – Basic principles (input-process-output, coding/decoding ...) of communication-technical transmission – Possibilities of analog and digital information transmission – Interaction of hardware and software – Programming of simple digital systems	×		×	×	
			6	1	3	5	4

states address the concepts Algorithms and Programming and Computing Systems. The concepts Data and Analysis and Impacts of Computing are addressed in three of the four states. The concept Networks and the Internet is addressed exclusively in HH. The other four states (BE, BB, MV, ST) can each be assigned to only one concept. In SN and SH, an extra learning area was created in the curriculum for informatics content. In the other six federal states, informatics content was integrated into thematically different learning areas. Mapping states with informatics content to the concepts of the CSTA K-12 CS Standards shows that topics that can be mapped to the concepts Computing Systems and Algorithms and Programming are most frequently assigned in six and five of the eight states, respectively. Content related to these concepts appears to be given the greatest importance. Content on the concept Networks and the Internet was integrated in only one of the states.

The compilation and examination of the current German primary school curricula have shown that the situation varies considerably with regard to the integration of informatics content in primary school. Compared to the CECE report from 2017 [22], in which none of the German states had yet integrated informatics content, 50% of the states now have informatics content included in their primary school curricula. However, 50% of these are exclusively attributable to only one of five concepts of the CSTA K-12 CS Standards [4]. Accordingly, one cannot speak of comprehensive informatics education in these cases. On the contrary, only one concept has been explicitly singled out in these places.

4   Interpretation of the Results

The recommendations on competencies for the primary level from the German Informatics Society aim at a broad view on informatics already in primary school [3]. The presented results show that this is implemented in only a few German states so far. Only in HH content on all five concepts of the CSTA K-12 CS Standards [4] has been integrated into the curricula. The most common concepts Computing Systems and Algorithms and Programming are those that are also prominent in the curricula of the UK [6], Sweden [11], and Lithuania [5]. In 2017, the concept Algorithms and Programming was already found in the strategy "Education in the Digital World" from KMK [13, p. 13], which explains its frequent occurrence. The fact that the concept of Network and the Internet is integrated as informatics content in only one of the German federal states may be related to the fact that in Germany there is a strong focus on the use of the internet and the associated dangers, and not yet on how the internet is structured and functions [13, p. 3, 13]. A look at the curricula selected as examples from Sweden, Lithuania and the UK shows that in Sweden and Lithuania the structure and functioning of networks and the internet are not explicitly addressed [5, p. 89] [11, p. 121 f], but in the UK they are addressed in key stage 2 [6]. At this point, there is great potential for development, since the use of the internet is dealt with in almost all current German primary school curricula and thus the linking of the technological background is an important further step. The situation is similar with dealing with one's own data and data protection. In this respect, the curricula have so far mainly dealt with dangers and instructions on how to deal with them in a critically competent manner, but the technological background, is not considered. If you pick up there, you can at least partially assign the concept Impact of Computing.

The federal states that have not yet integrated informatics content into their curricula can also start at various connecting points beyond this. When scanning the curricula, text passages were found that, although not explicitly informatics in nature, certainly offer opportunities for tying in informatics content. In some curricula (e.g. BY, HB), the function and use of everyday objects and learning how simple machines and devices work in schools and private households are addressed. Just like the topic of human inventions (e.g. NI) or working and living in the past and today, there are possibilities to simply choose informatics systems as a subject of investigation. Also, the topic of automation of certain activities, which used to be very laborious (for example, washing machine versus washing clothes by hand) lends itself (e.g. HB, MV). The acquisition and representation of data, which is anchored as a topic in a large part of the math curricula, was not evaluated as explicit informatics content in the context of this paper. What matters here is how the topic is addressed by the teacher. When implementing the acquisition and representation with digital media, the thematization of digitization, the transformation of analog into digital data, is obvious as informatics content.

The naming of informatics in superordinate sections of the curricula is already implemented in some federal states. Here, care should be taken to ensure that the naming is also reflected in the content of the learning areas, so that teachers have an orientation as to how this overarching goal is to be implemented in concrete terms. The expansion of the naming of informatics education as an overarching goal could be used to identify the concrete informatics content in the learning areas. For the teachers it would then be obvious that at this point the concrete functioning is to be questioned and the informatics background is to be highlighted. As a basis for this, for example, standards for informatics education in primary school could be supported at the state level, which would then subsequently be integrated into the curricula of the 16 federal states. Especially HH, NW, SN and SH can serve as role models and inspiration by integrating informatics content out of three or more concepts of the CSTA K-12 CS Standards. It should be emphasized that the integration of informatics content into the curricula alone is not sufficient to bring informatics education into primary schools. It is important that the content also finds its way into teacher training, as the teachers then have the task of teaching this content to the primary school students.

5   Discussion of the Research Methodology

Since the research was limited to a selection of 16 keywords for finding informatics content in the curricula, it is possible that not all existing informatics content was found. When using the term digital it became clear that the mentioning of the word pair digital media rather indicates media-forming aspects. In the cases in which the found content is about informatics at least one or more of the already used keywords are mentioned. Due to the partly very short mention of informatics content in the curricula, there is some room for interpretation. In the curriculum of the subject technology (SH), "understanding and explaining basic principles (...) of communication-technical transmission" and the application of "possibilities of analog and digital information transmission" [17, p. 16] are integrated. The term transmission could also be interpreted in the direction of the concept Networks and the Internet. Since neither the explicit treatment of networks nor internet was named, it was not assigned to this concept.

6   Outlook

Since a more informatics-oriented supplement to the strategy "Education in the Digital World" from KMK was published at the end of 2021 [14], and the German Rector's Conference also spoke out in favor of informatics education in all teacher training programs at the beginning of 2022 [9], it can be expected that the integration of informatics content into German primary school curricula will develop positively in the coming years. In order to make these developments visible, the current state (as for the situation of informatics teaching in secondary schools [20]) should be reviewed regularly. In addition, exchange with the state authorities would be useful to investigate further developments that are still in the planning stage.

References

1. Behörde für Schule und Berufsbildung Hamburg: Sachunterricht (2011)
2. Behörde für Schule und Berufsbildung Hamburg: Mathematik. Anlage zur Umsetzung der KMK-Strategie "Bildung in der digitalen Welt" (2020)
3. Best, A., et al.: Kompetenzen für informatische Bildung im Primarbereich (2019). https://dl.gi.de/bitstream/handle/20.500.12116/20121/61-GI-Empfehlung_Kompetenzen_informatische_Bildung_Primarbereich.pdf?sequence=1&isAllowed=y
4. Computer Science Teachers Association: CSTA K-12 Computer Science Standards (2017). http://www.csteachers.org/standards
5. Dagienė, V., Jevsikova, T., Stupurienė, G.: Introducing informatics in primary education: curriculum and teachers' perspectives. In: Pozdniakov, S.N., Dagienė, V. (eds.) ISSEP 2019. LNCS, vol. 11913, pp. 83–94. Springer, Cham (2019). https://doi.org/10.1007/978-3-030-33759-9_7
6. Department for Education: National curriculum in England: computing programmes of study (2013). https://www.gov.uk/government/publications/national-curriculum-in-england-computing-programmes-of-study/national-curriculum-in-england-computing-programmes-of-study
7. European Commission: Digital Education Action Plan 2021–2027 - Resetting education and training for the digital age. https://education.ec.europa.eu/sites/default/files/document-library-docs/deap-communication-sept2020_en.pdf

8. European Commission: Joint Research Centre. Reviewing computational thinking in compulsory education: state of play and practices from computing education (2022). https://data.europa.eu/doi/10.2760/126955
9. German Rector's Conference: Teacher education in a digital world. Resolution of the Senate of the HRK on 22 March 2022 (2022). https://www.hrk.de/resolutions-publications/resolutions/beschluss/detail/teacher-education-in-a-digital-world/
10. Gülbahar, Y., Ilkhan, M., Kilis, S., Arslan, O.: Informatics Education in Turkey. Commentarii informaticae didacticae: (CID) (6), 77–87 (2013). https://publishup.uni-potsdam.de/opus4-ubp/frontdoor/deliver/index/docId/6213/file/77_87_GAlbahar_etal.pdf
11. Heintz, F., Mannila, L., Nordén, L.Å., Parnes, P., Regnell, B.: Introducing programming and digital competence in Swedish K-9 education. In: Dagiene, V., Hellas, A. (eds.) ISSEP 2017. LNCS, vol. 10696, pp. 117–128. Springer, Cham (2017). https://doi.org/10.1007/978-3-319-71483-7_10
12. Jovanov, M., Stankov, E., Mihova, M., Ristov, S., Gusev, M.: Computing as a new compulsory subject in the Macedonian primary schools curriculum. In: 2016 IEEE Global Engineering Education Conference (EDUCON), pp. 680–685. IEEE, Abu Dhabi, April 2016. https://ieeexplore.ieee.org/document/7474623
13. Kultusministerkonferenz: Bildung in der digitalen Welt. Strategie der Kultusministerkonferenz (2016). https://www.kmk.org/fileadmin/Dateien/pdf/PresseUndAktuelles/2017/Strategie_neu_2017_datum_1.pdf
14. Kultusministerkonferenz: Die ergänzende Empfehlung zur Strategie "Bildung in der digitalen Welt" (2021). https://www.kmk.org/fileadmin/veroeffentlichungen_beschluesse/2021/2021_12_09-Lehren-und-Lernen-Digi.pdf
15. Landesinstitut für Schule und Medien Berlin-Brandenburg: Sachunterricht (2015)
16. Ministerium für Bildung und Kindertagesförderung Mecklenburg-Vorpommern: Sachunterricht (2020)
17. Ministerium für Bildung, Wissenschaft und Kultur des Landes Schleswig-Holstein: Technik (2021)
18. Ministerium für Schule und Bildung Nordrhein-Westfalen: Lehrpläne für die primarstufe (2021)
19. Niedersächsische Landesinstitut für schulische Qualitätsentwicklung (NLQ): Mathematik (2017)
20. Schwarz, R., Hellmig, L., Friedrich, S.: Informatik-Monitor (2021). https://informatik-monitor.de/
21. Staatsministerium für Kultus Freistaat Sachsen: Werken (2019)
22. The Committee on European Computing Education (CECE): Informatics Education in Europe: Are We All In The Same Boat? (2017). https://doi.org/10.1145/3106077
23. The Informatics for All Coalition: Educating People for the Digital Age (2020). https://www.informaticsforall.org/wp-content/uploads/2020/07/Informatics-for-All-position-paper.pdf

A Tool to Create and Conduct Custom Assessments in Turtle Graphics

Jeremy Marbach[1], Alexandra Maximova[1], and Jacqueline Staub[2(✉)]

[1] Department of Computer Science, ETH Zürich, Universitätstrasse 6,
8092 Zürich, Switzerland
`jmarbach@student.ethz.ch`, `alexandra.maximova@inf.ethz.ch`
[2] Fachbereich IV, University of Trier, Behringstraße 1, 54296 Trier, Germany
`staub@uni-trier.de`

Abstract. With the recent introduction of computer science in elementary school, teachers must monitor their students' progress in a subject known for both its creative and challenging nature. Assessing a vast number of diverse solutions is a time-taking challenge and, without the help of automation, some teachers may be tempted to resort to traditional assessment techniques that are easier to verify but do not provide the same possibilities for constructionist learning. We present a tool to create and conduct custom programming assessments in turtle graphics. Solutions are verified by pixel-wise comparison of student and sample solutions while also considering constraints on the set of possible solutions. The tool has been deployed in the XLogoOnline programming environment and we are planning to further analyze student performance in practice.

Keywords: CS education · K–6 · Programming · Assessment · Turtle graphics

1 Programming Assessment – A Teacher's Nightmare?

With various school reforms across the globe [2,8,10], computer science is finally paving its way into primary school education. The role of computer science for general education is considered to lie in the area of computational thinking, a skill that can be fostered (among others) through programming. Children as young as nine or ten years now have the chance to learn to program in school using didactically-enhanced learning environments and teaching methodologies [6,22]. At this age, students are typically introduced to basic programming concepts such as sequences and loops; once they are older they tackle also more abstract concepts such as procedures, parameters and variables [12,13].

Programming is known to be a creative activity which requires a high level of semantic accuracy [11,21]. A given task can usually be solved in numerous different ways [18] and quickly comprehending the different strategies novices may choose is highly demanding, even for experts [5]. Teachers without a strong

A. Bollin and G. Futschek (Eds.): ISSEP 2022, LNCS 13488, pp. 15–26, 2022.
https://doi.org/10.1007/978-3-031-15851-3_2

background in computer science (as they are still often found in primary and middle schools [15,19]), understandably, are challenged even more in judging whether a task is solved correctly or not, based on just a given program.

Formative and summative assessment is considered an essential tool [3,4,26] that allows teachers to observe their students' individual learning progress and analyze the teaching-learning process as a whole [9]. Usually, assessments are designed to quantify competencies in a "typical" work environment. In the context of programming classes, we seek for assessment techniques for the creative process of solving problems using computational thinking skills.

Implementing formative and summative assessment tools in the context of primary school programming is no easy task. Although programming classes offer the computer as an additional help with the power of automation, there are few widely adopted and well-established automatic assessment tools [16]. With few exceptions [7,17], most of the available tools neither consider the special requirements in primary schools nor their dedicated application domains (e.g. turtle graphics). Literature classifies assessment tools depending on whether they are student-centered, teacher-centered, or hybrid tools and whether they provide support for manual, semi-automatic, or automatic assessment [23].

This article presents an automatic and student-centered assessment tool for primary schools that provides the possibility to create custom exercise collections in turtle graphics geometry. The tool is integrated into the XLogoOnline programming interface [14,24] and hence gives the opportunity to assess learners in an environment they are familiar with. The tool is able to auto-verify student solutions to correctly make use of programming concepts such as sequences and loops, and more over, is able to verify whether extra-imposed conditions on the choice or number of commands is adhered to. The presented tool builds on an existing system [25] applicable to a younger age group covering simple navigation tasks instead of classical geometry tasks.

In Sect. 2, we give a short overview of the application domain turtle graphics as well as the curriculum that we use in our approach. Section 3 then shows a data structure for the representation of tasks and we discuss how student solutions can be verified automatically. Section 4 discusses the question of how the scope between "correct" and "incorrect" solutions could be accounted for before Sect. 5 finally draws some implications and possibilities for future work.

2   A Programming Curriculum for Grades 3–4

Over the past 15 years the Center for Computer Science Education (ABZ) and the Chair of Information Technology and Education at ETH Zurich have developed a spiral curriculum to teach computer science and programming at all school levels. Throughout the curriculum, turtle graphics is used to teach basic programming concepts. The corresponding programming environment XLogoOnline is developed by the Turtle Group Trier (TGT) and is currently used by almost 85 000 users per year.

2.1   The Idea Behind the Turtle

The turtle was introduced by Seymour Papert in the late 1960s as an *object to think with* [20]. Just like a person, an animal or a vehicle, the turtle is an object with a position in space and an orientation; it exists in both virtual and physical realizations. Using a personified object establishes an age-appropriate mental model of program execution and allows novices to identify themselves with the turtle. Such a target for synchronicity is considered helpful for novice programmers, assisting them in learning to communicate with the computer in a language that the machine "understands".

2.2   The Vocabulary

The turtle's mother tongue consists of only six parameterized *commands*. The two movement commands `forward` and `back` are used to steer the turtle forwards or backwards a given number of pixels. The two rotation commands `right` and `left` are used to turn the turtle by a given angle to the left or to the right. These four basic navigation commands are interpreted from the turtle's perspective and oftentimes have a two-letter abbreviation (i.e., `fd`, `bk`, `rt` and `lt`). In our block-based environment, the `setpencolor` command allows the pen color to be changed from initially black to another color from the color palette (i.e., black, white, green, blue, red and yellow). The last command `repeat`, finally, is the first control structure pupils encounter which allows them to describe repeating program behavior in shorter and more elegant programs using loops.

2.3   The Curriculum

A surprisingly large variety of geometric shapes can be created using this small instruction set; from simple patterns like squares and rectangles to more complex shapes like polygons, circles or mandalas. We present our curriculum as a suggested route to explore the task space, following five rough milestones.

1. **Exploring basic commands**
   In the beginning, only the four commands `forward`, `back`, `left` and `right` are used while pupils learn to navigate the turtle. In this stage, novices may execute one or few commands at a time to verify their logic along the way (see Fig. 1, task 1).
2. **Working with a restricted command set**
   Students not only learn to express themselves with the vocabulary available, but they also analyze the expressiveness of the language itself. They learn that the instruction set is redundant,i.e., with one movement and one rotation command, the other two basic commands can be substituted. To illustrate this point, teachers may prohibit some command to be used while solving a given exercise.
3. **Introducing colors**
   Colors are introduced once pupils feel confident with the four movement and

rotation commands and understand that there are several possible ways to draw the same shape (see Fig. 1, task 2). As before, students should discover that there are different approaches to solving the same problem by discussing and comparing their solutions with each other.

4. **Making use of the looping construct** `repeat`
   The ability to recognise repeating patterns and to express oneself using loops is a key milestone in pupils' programming development. However, due to the cognitive difficulty, some students might want to avoid loops unless they are explicitly challenged to apply the concept in their solution. Two ways of doing so are illustrated in Fig. 1 (tasks 3 and 4); once by phrasing the exercise statement in a specific way and once by choosing an exercise that invites students to use a loop (i.e., filled rectangles can be drawn tracing many lines right next to each other; drawing such a 100 by 100 square without a loop requires some 500 commands).

5. **Nested loops and sequences of loops**
   The last step of this curriculum is only used with advanced students: Based on their understanding of simple loops, students move on to sequences of loops and nested loops.

The curriculum is supported by an extensive exercise collection built into the XLogoOnline programming environment for grades 3 and 4 (children aged 9 and 10). Moreover, teachers (and students) have the possibility to create and share their own tasks. Figure 1 shows how the same geometric shape, a square, can be used repeatedly throughout the curriculum with increasing difficulty attached to it. The next section discusses how such tasks can be created using our tool and how their automatic verification process works.

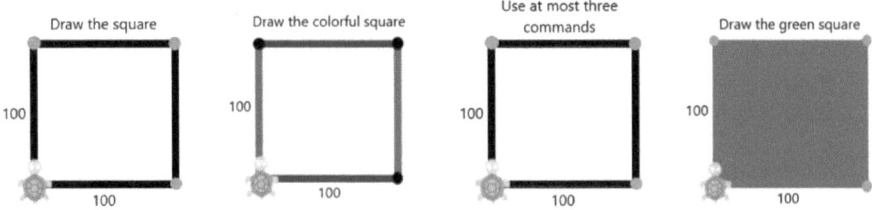

Fig. 1. Four tasks students encounter while working on our programming curriculum. From left to right: (1) a task that can be solved in any desired way (2) a task that requires colors, (3) and (4) are tasks that require the use of loops. (Color figure online)

3   Defining a Data Structure for Task Representation

All turtle graphics tasks used in our curriculum share the same fundamental structure. We present a data structure that distills these structural attributes and can be used to describe any turtle graphics task with and without colors. Later, we dive into the topic of solution verification and constraint handling.

3.1 Pixels and Lines – The Atomic Structures of the Turtle Universe

All geometric patterns mentioned before can be represented by lines. What is easy to explain for squares and stairs is, however, also true for filled rectangles and circles. The argument is that filled rectangles are in essence nothing else than a large number of lines directly adjacent to each other, while circles can be approximated by polygons with a large number of edges. What sounds like a gimmick has significant implications: No matter the shape, it can be drawn using just a number of simple line segments – a basic data structure therefore does not need anything in addition.

As a basic example, a square can be represented using just four *lines*. Each such line involves two coordinates (x_1, y_1) and (x_2, y_2), specifying the line's start- and endpoint. The example shown in Fig. 2 can be formally described as a set of four lines, each declared by two coordinates representing the respective start- and endpoints: $\{((0,0),(0,-100));((0,-100),(100,-100));((100,-100),(100,0));((100,0),(0,0))\}$.

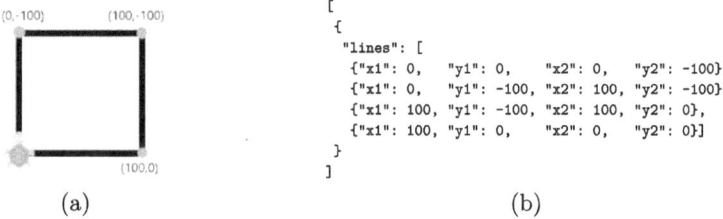

(a)   (b)

Fig. 2. (a) A square with four lines that connect the edge coordinates and (b) its corresponding representation using line segments

In constructionist manner, turtle graphics offers its users freedom in the implementation of the given geometric figure. Even with a simple figure like a square, there are numerous different solution programs, depending on the chosen order and direction in which lines are visited. While it is possible to allow freedom even in the choice of the turtle's starting point, specifying which coordinate the program must start from reduces uncertainties in the verification process [7]. Our implementation considers the coordinate $(0,0)$ always to be the turtle's implicit starting point as well as (at least) one of the line segments to start or end at this very coordinate. On top, the turtle's initial orientation is considered to head north in all cases. While the turtle's initial position and orientation are fixed, lines can have any arbitrary start and endpoint which is the minimum specification required to represent an exercise.

Up to this point, the world of XLogoOnline was perceived only as black and white. In reality, however, many interesting tasks can be phrased involving the concept of colors. How the described specification can be extended with colors and what implications arise from such a change will be covered next.

3.2   Introducing the Concept of Color

Turtle geometry shares some of its fundamental concepts and ideas with conventional geometry, i.e., the turtle universe can be understood as an arbitrarily large two-dimensional space in which various coordinates can be visited and connected. In contrast to conventional geometry, however, the otherwise infinitesimally small point coordinates are represented by discrete pixels which each have their own color. Any two coordinates that are connected by a sequence of consecutive pixels of the same color can be considered to be connected by a line.

In order to reflect the concept of line colors in the data structure, we extend each line object with an extra label col marking the line's color. Each line segment may have its own color as shown in Fig. 3 where the square example from before was extended with two red and two blue line segments.

The turtle universe provides space for large, colorful but also complex shapes. In order to provide immediate feedback independent of class size and task complexity, verification aught to be automated. The next section presents an approach of how automatic solution verification works.

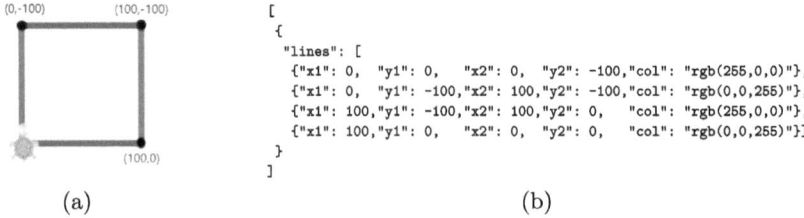

(a)   (b)

Fig. 3. A colored square (a) and its corresponding line segments (b) (Color figure online)

3.3   Preparing Student Programs for Verification

Upon pressing the $< PLAY >$-button, a given sequence of commands is transformed into an abstract syntax tree. This formal structure can be visited and used to extract the required line segment for a student solution. Figure 4 shows an example: the Logo program fd 100 setpc red repeat 4 [fd 100 rt 90] is represented as its corresponding abstract syntax tree is visited in a depth-first fashion. Each of the commands fd, bk, rt, lt, setpc and repeat mark individual inner nodes. Upon visiting the fd- or bk-nodes, a line object is generated and appended to an external data structure element. By default, line segments are colored black; once a setpc command is visited, however, the color flag of all subsequent line segments is adapted. The looping construct repeat, finally, has a special role: it re-executes its subtree block as often as requested, resulting in an implicit loop-unroll of line segments.

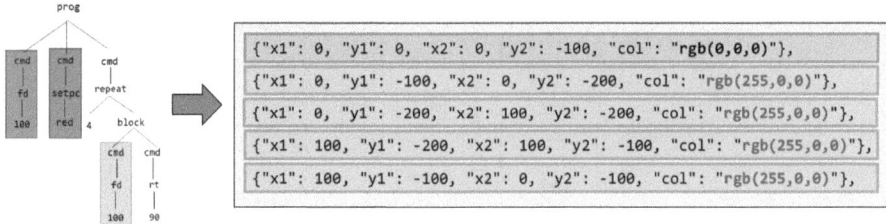

Fig. 4. Movement commands are converted into line segments while visiting the abstract syntax tree. This way, student programs are converted into line segments that can later be verified against. (Color figure online)

3.4 How to Verify the Correctness of Student Solutions

We define a *verifier* to be an algorithm that receives (i) a pre-processed *student solution* and (ii) a *sample solution* that need to visually match. In this context, we define the two terms student and sample solution as follows:

- **Student Solutions**, as declared in Sect. 3.3, are a sequence of line segments that capture the essential information about the visual effect of a given Logo program. For each execution attempt, the student program is transformed into a sequence of line segments and passed on for verification.
- **Sample Solutions** have the same structure as student solutions. In contrast to a student solution, however, sample solutions are defined ahead of time and remain static throughout the entire exercise.

In order to verify whether a student solution matches a given sample solution, both are converted into bitmaps and compared pixelwise. With the turtle's start position being fixed to the coordinate $(0, 0)$, both solutions can be aligned and compared easily. In order to avoid rejection due to aliasing, the comparison takes into account only those pixels with RGB-values matching exactly one of the pre-defined colors red, green, blue, yellow and black. White pixels in the student solution are not compared against the (potentially non-white) pixels in the sample solution in order not to reject partial solutions on the student side.

Despite careful comparison, this process may result in false negatives. How such occasions occur and a measure to combat it is presented in the next section.

3.5 The Problem of Comparison on a Pixel Level

A given graphical pattern, say a square, can often be created in various different ways. While both `repeat 4 [fd 100 rt 90]` and `repeat 4 [lt 90 bk 100]` produce the same visual result (a square), the underlying data structure differs in terms of line ordering (see Fig. 5). This seemingly insignificant difference causes problems once colors are introduced.

The order of lines determines the order in which lines are drawn onto the canvas (i.e., in a top-down fashion). In case of overlapping lines, such as the

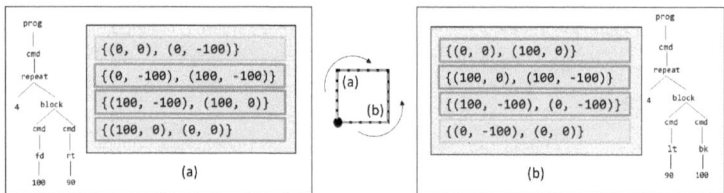

Fig. 5. Line segments for `repeat 4 [fd 100 rt 90]` and `repeat 4 [lt 90 bk 100]` are the same excepts for line ordering. The two parse trees create the same visual results, once (left) in a clockwise direction, once (right) counter-clockwise. (Color figure online)

edge of a square, minuscule pixel differences can cause seemingly-correct student solutions to be rejected. Figure 6 visualizes this point; although both the middle and the right square seem the same on a macroscopic level; their edge pixels differ due to line ordering (clockwise versus counter-clockwise). Seeing that sample solutions only capture one possible solution, they cannot be taken as ground truth down to the pixel-level.

To tackle this problem, verification can be extended with a *threshold*, allowing all solutions that match to a "high-enough" degree to be accepted. Finding a suitable threshold, however, has proven to be a challenge in its own right, as we will further discuss in Sect. 5.

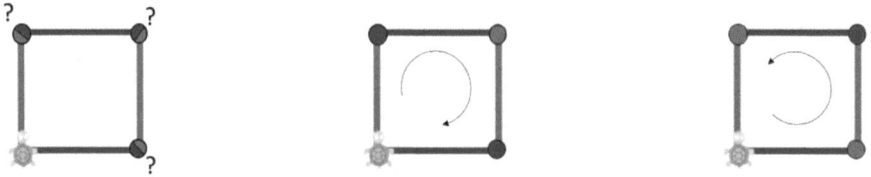

Fig. 6. Line ordering may cause pixel-differences in intersecting pixels.

3.6   Introducing Constraints

The question of how a solution was created is of crucial importance in constructionist programming classes. Educators have an interest not only in evaluating student solutions visually, but also in examining and evaluating their programmatic approach. To this end, we introduce the concept of *constraints*; a means to restrict the linguistic flexibility in student solutions.

As shown in Fig. 1, a simple shape like a square can be programmed in various different ways. While in the beginning, any working solution is good enough, more advanced students are expected to make use of programming concepts such as loops. Rather than promoting long and messy programs (e.g., instead of `fd`

100 rt 90 fd 100 rt 90 fd 100 rt 90 fd 100 rt 90 we want to enforce the more elegant and shorter alternative repeat 4 [fd 100 rt 90]) we want to ensure the looping construct is properly used and understood.

In order to enforce programmatic restrictions, we extended our data structure with an extra constraint field that allows the occurrence of fd, bk, rt, lt, setpc and repeat commands to be specified and captured in an automatically-verifiable formal task description. For each of these commands, the number of occurrences can be specified as an *upper bound, lower bound* or *exact count*. Listing 1.1 for instance, shows a task constraint for a square that is supposed to be drawn using *exactly* one fd-, one rt- and one repeat-command. There are other alternatives as to how this constraint can be expressed and, for instance, the total number of commands in a program (independent of the kind of command) could be used for the same purpose as well. This extension allows for more interesting and constraining exercise statements to be used, such as the third and fourth task in Fig. 1. Verifying student solutions can be performed using string analysis on a user program.

```
[
 { "lines":
    [{"x1":0,   "y1": 0,   "x2": 0,   "y2":-100, "col": ...},
     {"x1":0,   "y1":-100,"x2": 100,"y2":-100, "col": ...},
     {"x1":100,"y1":-100,"x2": 100,"y2": 0, "col": ...},
     {"x1":100,"y1": 0,   "x2": 0,   "y2": 0, "col": ...}]},
  { "constraints":
    [{"fd": {"type": "eq", "number": "1"},
     {"rt": {"type": "eq", "number": "1"},
     {"repeat": {"type": "eq", "number": "1"}]}
]
```

Listing 1.1. A task description that requires a square to be drawn using one fd-, one rt- and one repeat-command.

4   The Spectrum Between Right and Wrong

The presented verifier assesses the correctness of a solution based on pixel-comparison and constraints. However, mere correctness is only one of the aspects a teacher might want to take into account when assessing their pupils. In order to assess the expertise of pupils, both the overall time per task and number of incorrect attempts may be taken into consideration. Our tool computes the *score* of a solution taking into account all three aspects: correctness, time, and number of attempts. Our score function is inspired by the one used in the programming competition platform Topcoder [1].

Two variables can be set to compute the overall *score* s_i per pupil: T is the overall time provided to solve all tasks and p_i is the maximum number of points awarded per exercise. When a task is being solved. The time t_i and the number of attempts q_i until a correct solution is found, are registered resulting in the following score function that is subject to $a + b + c = 1 \wedge a, b, c \geq 0$:

$$s_i = p_i \left(a + \frac{bT^2}{10t_i^2 + T^2} + \frac{c}{q_i} \right),$$

The maximum number of points p_i that can be awarded for a task is thus subdivided into three parts: The a-portion of the points is awarded for correctness as defined in the previous section, no matter the time taken or the number of attempts. The b-portion of the points are subject to non-linear decay depending on the time t_i taken to solve a task correctly. The c-portion of the points decay linearly with the number of wrong attempts. We manually decided for $a = 0.7$, $b = 0.2$ and $c = 0.1$.

5   Conclusion

Computer science in many ways holds a special position in schools: In addition to the struggle of carrying the constructivist view of a long misunderstood discipline into their class rooms, teachers must gain the relevant content knowledge (i.e. programming skills) by themselves and figure out how to assess their students' learning. Despite its exacting nature, programming oftentimes allows for numerous different solutions that may vary greatly on both a syntactic and semantic level. Assessing a large number of diverse student solutions can be a tedious and time-consuming endeavor.

We presented a tool that allows teachers, potentially without much background in computer science, to easily design custom assessments in turtle graphics. The presented tool first verifies whether a student solution visually matches a given sample solution, as discussed in Sect. 3.4, and then ensures that any imposed constraints on the program-level are met as well. The tool has been integrated into the XLogoOnline programming environment that is widely used in primary and lower secondary schools across the globe. This work provides a stable basis for further analyzing student solutions on a primarily semantic, rather than syntactic, level. There are, however, also two noteworthy limitations:

1. While the presented data structure allows geometric shapes to be described formally, the representation is not optimized for space efficiency. One fundamental problem is that turtle graphics is not only used for simple geometric figures with few lines, but also for tasks that use a large number of line segments. For instance, circles can be approximated as polygons with a large number of lines and filled rectangles are constructed using a large number of adjacent line segments. Such figures take up an undesirable amount of memory, which can be problematic in terms of both storage and data processing.
2. A second yet unresolved question is how to find a threshold value that is universally suitable. As discussed in Sect. 3.5, it is not possible to simply check for an exact pixel match due to layering problems that results in pixel differences. It is difficult to find a tight threshold that works in all cases: the case of a colored cross cannot be handled the same as a four-colored

square which again cannot be handled the same as a filled-in square with an different-colored space in the middle (Fig. 7). Finding a suitable threshold is relevant; a too large threshold accepts incorrect solutions whereas a too small one might reject correct solutions.

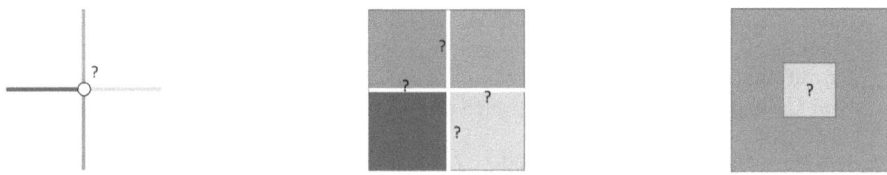

Fig. 7. Difficulty of finding a suitable threshold: (left) one pixel of uncertainty, (middle) two lines of uncertainty, (right) an area of uncertainty. (Color figure online)

References

1. Ratings in topcoder. https://community.topcoder.com/tc?module=Static&d1=help&d2=ratedEvent. Accessed 26 Jan 2022
2. Bell, T., Andreae, P., Lambert, L.: Computer science in New Zealand high schools. In: Proceedings of the Twelfth Australasian Conference on Computing Education, vol. 103, pp. 15–22 (2010)
3. Black, P., Wiliam, D.: Developing the theory of formative assessment. Educ. Assess. Eval. Accountability (Formerly J. Pers. Eval. Educ.) **21**(1), 5–31 (2009)
4. Boston, C.: The concept of formative assessment. Pract. Assess. Res. Eval. **8**(1), 9 (2002)
5. Brown, N.C., Altadmri, A.: Investigating novice programming mistakes: educator beliefs vs. student data. In: Proceedings of the Tenth Annual Conference on International Computing Education Research, pp. 43–50 (2014)
6. Dagiene, V., Hromkovic, J., Lacher, R.: Designing informatics curriculum for K-12 education: from concepts to implementations. Inf. Educ. **20**(3), 333–360 (2021). https://doi.org/10.15388/infedu.2021.22
7. Eschbach, D.: A computer-based examination system for XLogoOnline. Master's thesis, ETH Zurich (2019)
8. Fowler, B., Vegas, E.: How England implemented its computer science education program. Center for Universal Education at The Brookings Institution (2021)
9. Garrison, C., Ehringhaus, M.: Formative and summative assessments in the classroom (2007)
10. Goode, J., Skorodinsky, M., Hubbard, J., Hook, J.: Computer science for equity: teacher education, agency, and statewide reform. Front. Educ. **4**, 162 (2020)
11. Hristova, M., Misra, A., Rutter, M., Mercuri, R.: Identifying and correcting Java programming errors for introductory computer science students. ACM SIGCSE Bull. **35**(1), 153–156 (2003)
12. Hromkovič, J.: Einfach informatik 5/6: Programmieren. primarstufe. begleitband. Einfach Informatik (2019)

13. Hromkovič, J., Kohn, T.: Einfach informatik 7–9: Programmieren. sekundarstufe i. Einfach Informatik (2018)
14. Hromkovič, J., Serafini, G., Staub, J.: XLogoOnline: a single-page, browser-based programming environment for schools aiming at reducing cognitive load on pupils. In: Dagiene, V., Hellas, A. (eds.) ISSEP 2017. LNCS, vol. 10696, pp. 219–231. Springer, Cham (2017). https://doi.org/10.1007/978-3-319-71483-7_18
15. Hunt, N.P., Bohlin, R.M.: Teacher education students' attitudes toward using computers. J. Res. Comput. Educ. **25**(4), 487–497 (1993)
16. Ihantola, P., Ahoniemi, T., Karavirta, V., Seppälä, O.: Review of recent systems for automatic assessment of programming assignments. In: Proceedings of the 10th Koli Calling International Conference on Computing Education Research, pp. 86–93 (2010)
17. Kurvinen, E., Lindén, R., Rajala, T., Kaila, E., Laakso, M.J., Salakoski, T.: Computer-assisted learning in primary school mathematics using ViLLE education tool. In: Proceedings of the 12th Koli Calling International Conference on Computing Education Research, pp. 39–46 (2012)
18. Menta, R., Pedrocchi, S., Staub, J., Weibel, D.: Implementing a reverse debugger for logo. In: Pozdniakov, S.N., Dagienė, V. (eds.) ISSEP 2019. LNCS, vol. 11913, pp. 107–119. Springer, Cham (2019). https://doi.org/10.1007/978-3-030-33759-9_9
19. Mozelius, P., Ulfenborg, M., Persson, N.: Teacher attitudes towards the integration of programming in middle school mathematics. In: INTED 2019, pp. 701–706 (2019)
20. Papert, S.: Mindstorms: Children, Computers, and Powerful Ideas. Basic Books, New York (1980)
21. Pea, R.D.: Logo programming and problem solving (1987)
22. Pérez-Marín, D., Hijón-Neira, R., Martín-Lope, M.: A methodology proposal based on metaphors to teach programming to children. IEEE Revista Iberoamericana de tecnologias del aprendizaje **13**(1), 46–53 (2018)
23. Souza, D.M., Felizardo, K.R., Barbosa, E.F.: A systematic literature review of assessment tools for programming assignments. In: 2016 IEEE 29th International Conference on Software Engineering Education and Training (CSEET), pp. 147–156. IEEE (2016)
24. Staub, J.: xLogo online-a web-based programming IDE for Logo. Master's thesis, ETH Zürich (2016)
25. Staub, J., Chothia, Z., Schrempp, L., Wacker, P.: Encouraging task creation among programming teachers in primary schools. In: Barendsen, E., Chytas, C. (eds.) ISSEP 2021. LNCS, vol. 13057, pp. 135–146. Springer, Cham (2021). https://doi.org/10.1007/978-3-030-90228-5_11
26. Taras, M.: Assessment-summative and formative-some theoretical reflections. Br. J. Educ. Stud. **53**(4), 466–478 (2005)

Informatics at Primary Education: Teachers' Motivation and Barriers in Lithuania and Turkey

Gabrielė Stupurienė[1]([✉]) [iD] and Yasemin Gülbahar[2] [iD]

[1] Institute of Educational Sciences, Vilnius University, Vilnius, Lithuania
`gabriele.stupuriene@mif.vu.lt`
[2] Department of Computer Education and Instructional Technologies, Ankara University, Ankara, Turkey
`gulbahar@ankara.edu.tr`

Abstract. Scientific research shows that the role of teachers is crucial in the integration of informatics at all educational stages. Besides some other barriers, teachers' lack of knowledge of informatics is believed to be a reason why educators are reluctant to get involved in informatics teaching. Hence, this study focused on the factors influencing primary school teachers' motivation and possible barriers for integrating informatics in lessons. The data were collected from an accessible sample by conducting semi-structured interviews with primary school teachers in Lithuania and Turkey in 2022. This paper also presents an overview of informatics primary education curricula in both countries. Results showed that despite different curriculum conditions and sociodemographic issues in two countries, primary school teachers' factors of motivation and barriers are the same. Motivation depends on teachers' willingness to acquire new digital competences, to go beyond being a regular teacher and to innovate in their school. Teachers understand the importance of keeping their students up-to-date with the newest educational technologies and CT competences, however they face serious challenges to overcome. We sum up with a discussion comparing curricula and teachers' attitudes in both countries and conclude with some insights and recommendations.

Keywords: Computers science education · Informatics education · Computational thinking · Primary education · Primary school teachers · Motivation and barriers · Teacher training

1 Introduction

With the growing influence of new technologies, schools are not ready to be active participants in 21-century challenges. Computational Thinking (CT) has achieved the status of essential 21st-century skill and is now included in school curricula all over the world, although there is no consensus on how to define it [15]. From an educational and curricular perspective, computational thinking

A. Bollin and G. Futschek (Eds.): ISSEP 2022, LNCS 13488, pp. 27–39, 2022.
https://doi.org/10.1007/978-3-031-15851-3_3

has been recognized for developing knowledge and understanding of concepts in CS as well as for significant contribution to general-purpose problem-solving skills [12]. Some of the definitions are superficial and have caused confusion among teachers who don't understand how to teach basic computing and how to assess student progress and teachers are asking for clarification [5]. Despite the recent widespread acceptance of computational thinking, there are some gaps and uncertainties in teacher training and professional development to understand the aims and intentions of computational thinking (CT) education [7]. As stated by Bocconi et al. (2022) [1], teachers play an important role in the successful integration of informatics and CT at any school level. Zhang et al. (2020) [17] specified a lack of programming and CT knowledge as a reason why teachers do not include CT perspectives in their practice. Providing teachers with the necessary instructional support, which will guide the teaching process in their classroom, plays an important role in the implementation of the curriculum [14]. In the face of rapid technological change, teachers in particular are expected to be lifelong learners with the latest knowledge and skills, as well as the ability to replace education with these new and innovative tools and technologies [8]. But the problem is that with the aging of teachers, not all of them are ready to be active learners and be ready to improve their limited knowledge of CT and informatics and build capacity in fundamental CT concepts and pedagogies. Related to this topic, Jocius et al. (2020) [13], proposed the 3C (the code, connect, and create) professional development model which was designed to support teachers while infusing CT into their classrooms. Results showed that this professional development model supported the integration process, and increased teachers' self-efficacy and beliefs regarding CT integration into disciplinary content. Consequently, it is very important to take into account teachers' self-efficacy as one of the motivating factors. In the study about informatics and CT in primary education in 2019, 27 interviewed countries have already introduced an informatics curriculum for primary education, 11 countries were under active development of a new curriculum and there were no curricula for informatics in primary education in 14 countries. The data collected showed that informatics is taught in some form in primary education in the majority of countries surveyed (83%). However, the level of implementation of informatics varies widely [3]. Hence, integration of informatics and teacher training are two important topics for the world where serious discussions are continuing by growing. Besides changing their curriculum to include informatics concepts and CT processes, most countries have already started to discuss whether to integrate artificial intelligence in the curriculum or not [16].

Based on these facts, the research presented in this paper has two goals. The first one is to overview two countries' informatics primary school curricula in order to have a background and perception. The second one is to identify influencing factors of motivation of teachers to integrate informatics and CT in primary education. To find out key challenges that teachers in two countries are facing.

1.1   Informatics in Primary Education in Lithuania

Lithuania has a long-standing tradition (from 1986) in informatics education, however not in primary education (ages 7–11). In 2019, Lithuania's Ministry of Education, Science and Sport approved the Guidelines for Updating the General Curriculum Framework [6]. This foresees a new informatics subject to replace the current Information Technologies subject, with the introduction starting from primary school. Proposed version of the informatics curriculum framework for primary schools (grades 1–4) includes these six areas [4]:

- Digital content: essential skills of working with digital devices; managing textual, graphical, numeric, visual, audial information; information visualization and presentation; digital content creation.
- Algorithms and programming: solving problems: algorithm, action control commands (sequencing, branching, looping), programming in a visual programming environment for children.
- Problem solving: essential technical and technological skills for working with digital devices: solving technical problems, evaluating and identifying suitable technologies for the selected problem, and creative use of technologies.
- Data and information: working with data skills: problem analysis, data collection, sorting, search and data management, content quality evaluation.
- Virtual communication: social skills in a virtual environment: continuous learning, e-learning, communication via email, chats, social networks, sharing, collaboration, and reflection.
- Safety: digital safety, safe work with digital devices; ethics and copyright issues of information processing and usage; safety, ethics, and copyright issues in virtual communication.

In 2020, one hundred primary schools started to pilot the proposed informatics curriculum. The pilot targeted the development of learning resources and textbooks, as well as teacher training. The full-scale implementation of the informatics curriculum commences in 2023. While this new primary education Informatics curriculum covers most of the major CT components, it is up to individual schools (and their teachers) to decide how and in which subjects to integrate the informatics components they choose to address.

1.2   Informatics at Primary Education in Turkey

The first "Computer" course in K-12 was added as an elective course in 1997 with the decision of the Head Council of Education and Morality of the Ministry of National Education (MoNE). In this context, the course has been taught for 1–5 years, 1–2 h a week, starting from the 4th grade. Since the curriculum is designed in a spiral structure, students can choose this course starting from any class. This course was expanded to cover grades 1–8 later, and the choice of the course was left to the teachers' board. The course has been named differently, weekly hours and grades are changed from time to time besides revisions in the curriculum. In 2012, the name of the course was changed to "Information

Technologies and Software" and the course became compulsory in the 5th and 6th grades. Previously focusing on ICT concepts, the curriculum of the course has also been changed including "problem solving and programming". Lastly, in 2016, the curriculum of the courses for primary education has been slightly revised to include "Ethics" whereas a more important change in grades for 9th and 10th grades happened. For high school the course is named "Computer Science" and the curriculum shifted from teaching ICT to teaching "problem solving and programming" concepts. In science schools and social sciences schools, this course was made compulsory [9].

The whole curriculum from 1st grade to 10th grade is framed around basic topics with possible differences in sub-topics and difficulty in content. Hence basics were as follows:

- Information and Communication Technologies: the effect and importance of technology in terms of society, computer systems, file management etc.
- Ethics and Security: ethical values, digital citizenship, privacy and security, copyright and licensing etc.
- Communication, Search and Collaboration: computer networks, searching the net, communication tools, social media etc.
- Product Creation: developing documents and presentations, spreadsheets, using audio and visual tools, creating animations, 3D design etc.
- Problem Solving and Programming: problem solving approaches and tools, programming (block-based, text-based, physical, mobile, and web-based depending on the age level)

The curriculum is implemented in schools since then devoting the autumn semester to the first four headings and the spring semester only to the "Problem Solving and Programming" heading. Although no separate courses were announced for 1–4th grades in primary schools, MoNE delivered the curriculum (mostly based on CT) supported with books and activities and advised classroom teachers to deliver the content in free activity hours which is weekly 4 hours.

There were also many other initiatives for spreading coding around the country like makers and coding clubs for students, training opportunities provided to teachers and students with the support of governorships etc. However, due to some challenges like technological infrastructure and lack of knowledge of teachers, the implementation has not reached its full potential yet.

1.3   Summary of Two Countries Situation

As a summary, the policy documents overview highlights the fact that currently, both countries are not teaching informatics at the primary level in 1–4th grades. Turkey advised doing so by providing curriculum and support materials in 2016 whereas Lithuania decided to provide a compulsory course to integrate from grades 1–4 starting in the year 2023. The idea of integration should be at least 30% of the time or separate lessons one time per week from 2023. In 2019, 51% of primary teachers in Lithuania were at least 50 years old, and there was no systematic training, only some public or private initiatives (not free of charge). Starting with the F@tih project in 2011, Turkey, having quite a young generation of computing teachers, not only started to donate schools with the necessary technology and infrastructure but also started to provide continuous and planned teacher training activities in the form of training the trainers. Hence, having diverse implementations and situations, this paper aims to provide different insights and underline major and common problems together with possible suggestions.

2   Research Methodology

The aim of this research study is to reveal the current situation of teaching informatics in two countries, namely Lithuania and Turkey, in order to provide different perspectives on the same phenomena. For this purpose, qualitative analysis was carried out to explore the phenomena in depth. An overview of policy documents and curricula was used as a primary source in order to have an insight of both countries' intentions in teaching informatics. This study was planned while a Lithuanian researcher was doing an internship at a Turkish university, so there was an opportunity for collaboration. Moreover, accessible teachers were interviewed to gather varying experiences and thoughts.

Semi-structured interviews were conducted from January to May 2022. From Lithuania 8 primary school teachers and from Turkey 7 primary school teachers were interviewed about the integration of informatics into curricula. Respondents were the only ones who volunteered to take part in the research. Some of the interviews were done online through video conferencing software and some of them were carried out face-to-face. The interviews lasted about 40-60 minutes.

The average age of respondents in Lithuania was 48,5 years, with 6 out of 8 teachers with more than 20 years experience in primary education, one teacher with up to 15 years experience, and one teacher with up to 5 years experience in primary education. The average age of respondents in Turkey was 37 where 3 out of 7 had more than 20 years of experience whereas 4 other teachers had less than 7 years of experience. Age and experience have a great effect on informatics teaching since we are living in a digital age and transforming with technology. Hence having a diverse sample for this study, despite the small number of participants is important to provide a broader insight. Semi-structured interviews were conducted according to the Interview Guide, which has been designed based on literature review and the experience of researchers. It was validated by five experts for different areas (psychologists, educologist, and sociologists). The Interview Guide consisted of 17 questions.

The policy documents and curricula were analyzed using "content analysis" whereas interviews were analyzed through "inductive coding" in order to reveal emerging themes. Hence, based on content and inductive analysis of the policy documents and the interviews, findings are revealed and interpreted to provide a deep insight to the problem.

3   Results

In this section, the analysis of the data will be presented separately for each country to reveal possible differences from different perspectives.

3.1   Case of Lithuania

In order to have a broader view of Lithuania, 8 teachers were selected from 3 state schools (N = 6) and 2 private schools (N = 2) in 4 regions. Four interviews were done face-to-face and four by using an online video-conferencing tool. The analysis of interviews showed that there are several practices regarding informatics integration in schools and it is closely related to teachers' competencies and understanding of informatics but discouraged by a lack of tools and methodology. Three teachers have a very clear view on how to integrate informatics or even to teach as a separate subject; where four teachers have some misunderstanding about informatics as a subject and differences between digital competencies or lack of methodology on how to integrate and one teacher was really in doubtful situations with even technologies.

Integration of Informatics. Informatics is integrated or even is taught as a separate subject, not in all private and state schools and this mostly depends on teachers' competencies, schools administration, and infrastructure. Two experienced teachers from the same state school are integrating informatics and CT and have a separate lesson one time per week together with an IT teacher. They mentioned that what they do in a separate lesson sticks with the child more. Both of them think that it is useful to teach informatics from an early age because the skill has to grow: "In this world, it is probably no longer an option to leave technologies behind. The earlier students start to grasp all this and to further their knowledge the better." Teachers, who integrate informatics and CT use plugged and unplugged informatics activities. They use a variety of tools like Blue bot, Logo, Scratch, ScratchJr, Scottie Go!, Eduten, WordBall, Liveworksheets.com, Nearpod, StoryJump, Cospaces, Bebras cards. Some schools have a separate IT lab for primary school students. All schools also provide coding clubs/robotics which is an informal after-school activity for students.

Source of Motivation. Interviewed teachers think that the interest in technologies depends on themselves first. If teachers are not interested, it will be difficult to integrate informatics. There are schools where teachers don't want to be behind other colleagues, so they go and learn. Most teachers' motivation is curiosity, but some of them (up to 50 years) catch themselves thinking that they are probably in a generation where they don't learn as quickly and can't keep up with the young anymore.

Professional Development. All 8 teachers took part in the Teachers Lead Tech program, which is a paid private initiative for creativity in informatics and technology. The goal of this program is to patiently and adequately help prepare primary school teachers from the very beginning. For inexperienced teachers, this is a good opportunity to start to get acquainted with technologies, but experienced teachers lack progress because sometimes they know more than in this program. There are also some courses/seminars for free from universities and other educational institutions in order to provide new tools or methodology. All teachers have an opportunity to be part of the Methodology group of primary education in their region. An older teacher from state schools likes that she is involved in such group meetings, because together with other teachers they analyze new tools and activities, discuss, and reveal both their weaknesses and strengths. This teacher is very important to have a very good collaboration with colleagues in school and not be afraid to show that she doesn't know something.

The Role of Administration. Principals in one state and two private schools invest in technologies, not because it is fashionable, but because they see the benefits in it. They are very accepting, both interested and capable. In these three schools, students also have a separate lesson on informatics in the IT lab. In schools where integration of informatics is in its initial state, school principals and administration are open to helping teachers as much as they can but financial, innovation, and management issues are barriers. In schools where there is no integration of informatics despite teachers' understanding of informatics, but lack of pedagogical knowledge on how to integrate unplugged activities, the principal is open, but the school is in a rural and not rich area and they have very limited financial resources.

Parent Support. In all 5 schools, parents are helpful. But teachers also mentioned that they need to invest some time to talk with parents about ethics, safety, and data protection because not all parents are well educated to use technologies in a safe way. In one state school parents from two classes provide personal computers for their children to be used at school. It lets 2 teachers of this school integrate informatics and CT whatever they need.

Challenges. The challenges more experienced teachers have faced are that there is no systematic approach, and teachers are distracted. One young teacher from a private school misses clarity on how to integrate and what because now she is looking for herself on how to do it and wasting her time. Some teachers argue that informatics becomes a much more complex subject to integrate into primary school and requires a lot of effort compared to other subjects. They would like to know the key, how teachers should teach children in the best way and understand that instead of playing games they need to show students the real informatics, which is actually becoming relevant for that young child. All interviewed teachers (even older ones) argue that they can't ignore technology: "Children are born with technology, it would be a sin if they didn't introduce informatics in primary school". But overall there is a lot of dissatisfaction from older teachers in Lithuania for renewed curricula which include informatics. One teacher from primary school who partly integrated informatics thinks the biggest barrier for teachers is fear: "Here's a small child sitting down at a computer and he does everything right away. And we have fears: I can't do it, I won't succeed." In one state school, 2 teachers have an understanding of informatics but are not integrating, because they don't know how to do this without computers and other tools. It means that they don't know how informatics can be integrated by unplugged activities. One teacher from a state school has very low digital competence and she is worrying about technologies in general. She agrees that their students know more than she does but is trying to be open to any changes in curriculum. One more challenge is that a large part of certain platforms and applications require very long logins, especially if it is for children and this frustrates many teachers.

Suggestions. Recommendations from more experienced teachers would be that primary school teachers shouldn't be in charge of computers/tablets and other physical tools. Especially for those teachers who are still struggling with computers, they need help from an IT teacher/technical person, at least for the first year. Teacher's book/good practices material or practical training that clearly states what to teach from informatics and CT and how to do this would really be a big boost for most teachers. Also, teachers mentioned that they would like live training (not online), to try to learn by doing and understand practically, because the theory is one. Also to get supporting continuous learning and not one-off courses.

3.2 Case of Turkey

The situation in Turkey shows a great variety between state and private schools. For this reason, primary school teachers were selected both from the state (N = 4) and private (N = 3) schools for this study. Hence, the data will be presented by mentioning this gap wherever relevant. State school teachers were from Istanbul whereas private school teachers were from Ankara. That is why online interviews were held with teachers from Istanbul and face-to-face meetings were planned for teachers in Ankara.

Integration of Informatics. Like the majority of the private schools in Turkey, the teachers working at private schools mentioned that they have a separate informatics course starting from the 1st grade whereas there were no separate courses for state schools. Although MoNE prepared and published curriculum, activities, and even books for grades 1 to 4, only possible implementation can be done in free activity hours which is only 4 h a week. State school teachers mentioned they are teaching coding and using CS unplugged activities in these hours. All schools can also provide coding clubs which is an informal after-school activity for students. Since private schools have their separate course and curriculum they state that they teach both ICT, computing, and informatics at the same time. Some mentioned content is about keyboard usage, input-output concept, use of block-based programming (code.org, Scratch, kodu.org), three-dimensional design, use of programming tools, etc. State schools teachers stated that they are trying to teach similar concepts like coding-pixel painting and CS unplugged activities, using weekly free activity time to invite an expert from a private organization, they prefer gamified teaching of coding, especially in grades 2–3, and using Web 2.0 tools most of the time. Due to the curriculum, they stated that they have difficulties integrating informatics into 4th grade.

They also mentioned that they also integrate relevant concepts into Turkish, and Mathematics courses.

Assessment of Informatics Concepts. In terms of assessment state school teachers mentioned they evaluate their students based on learning outcomes and criteria defined through practices, presentations, and the use of rubrics. However, they also mentioned that since there is no official course, no need to do assessments and students are demotivated due to not being graded on their performance. Meanwhile, private school teachers mentioned that students make presentations, they use a paid learning environment for teaching and assessment through a QR Code approach, and they also assess their students' algorithmic approach and products they developed through physical programming. Moreover, students have their own blogs starting from 1st grade to 4th grade, which they use as an e-portfolio for evaluation. Private schools also integrate computing into math and social studies courses by using the STEAM approach from time to time. They also teach ethics in 1–3 grades besides IT topics.

Hence, while private schools provide a separate courses for students, state school teachers are using free hours and integrating into different disciplines. They integrate computational thinking algorithms into Turkish with "instructions" and into Math with "patterns". They also mentioned that implementation takes too much time, even in the breaks they have to work, and teaching both theories and making them practice requires more time. Luckily, students are open to change, open to learning, and ready to go digital and their self-esteem increases due to learning this kind of knowledge.

Source of Motivation. Private school teachers mention that their motivation comes from the willingness to integrate technology, so they are using Web 2.0 tools and interactive learning environments for teaching. They want to make students ready for real life so they teach technology through practice. They are also using an international Cambridge curriculum and preventing multiple choice until 4th grade. On the other hand, state school teachers' motivation comes from their loyalty to their profession and dedication to work. Teachers wish to diffuse innovations, so they have the self-control to be donated with up-to-date knowledge which in the end makes them get respect from their students. They want to be real role models for their students and feel that their behavior is valued by their students.

Professional Development. Thus, for professional development, all teachers are in search of appropriate opportunities. State school teachers are attending professional development activities with the support of parents, they are engaging in projects, and attend workshops and training provided by MoNE. They mention that university support is so important since schools are living labs for implementation. Private school teachers are also mentioned in similar activities where all of them are supported by their administration. They are attending workshops and training provided by their own academy besides participating in various teacher training activities, watching videos, and reading course notes.

The Role of Administration. It is obvious how important the administrators' role here is. While private school teachers mention that their administration supports projects, looks for innovations, and values this leadership, state school teachers are complaining about the lack of this technological leadership in their schools. All the teachers agreed on "school leaders' being also a technological leader is so important". Thus, they mentioned that school principals' have to take risks since their support is so important.

Parent Support. Likewise in all other questions, the situation of parent support also showed diversity between private and state schools. Parents in private schools are more supportive of innovations and the integration of informatics, they provide great support and even demand for training. Thus, they look for after-school makers and monitor the homework of their children. Even parents are provided training for secure and healthy use of technology to support themselves and their children. On the other hand in state schools, well-educated parents provide more support, while others don't care. Parents value concrete outcomes and for them, students' wellness is more important.

Challenges. Surprisingly, private schools are facing almost no challenges due to their administrations' points of view. Thus, we can conclude that they are successfully integrating informatics in primary education as a separate subject, through the STEAM approach by using constructivist approaches like problem-based learning. Unfortunately, state schools underline several obstacles they face for effective integration of informatics as follows:

- Lack of technical infrastructure, out-of-date technologies;
- No computing teachers in the first two grades due to system change 4 + 4;
- Overcrowded classes (20–25 students, should be at most 16);
- Resistance from parents;
- Lack of time.

Suggestions. As expected most of the suggestions are made by state school teachers. They mentioned that have to teach "data" concepts and "algorithms" through variables, loops, and constraints through different approaches. They mentioned logic and philosophy should be integrated starting from 1st grade in schools and thinking skills should be focused on so that children have the chance for improving their higher-order thinking skills. Informatics courses should be compulsory starting from the first grade and content should be culture-sensitive. Interdisciplinary teaching of informatics concepts should be planned and should aim for permanent learning.

4   Discussions and Conclusion

Following the purpose of the study, the policy documents and curricula about informatics in primary education have been overviewed and teachers are interviewed for their opinions and insight. Many countries have already been involved in the process of updating their curricula for informatics education at primary school [3,4]. But most countries are struggling with similar obstacles.

Integration of informatics in all countries, like in these two, strongly depends on teachers' experiences and attitudes toward the idea. Hence, although teacher training is one of the most important steps in achieving the goal of integration of informatics, even it is not sufficient if the teacher is not in favor of doing it or if the teacher doesn't value and believe its importance. This fact is true for both the countries, but Turkey has already started integrating informatics earlier in primary education, especially in most private schools and some state schools [10]. Although the percentage is small for state schools now, more teachers every day are provided necessary training both face-to-face and online through the delivery of instructional materials and activities. Hence, teacher training can be the most important requirement for effective integration [2,11].

Teachers of both countries stated that they cannot effectively teach informatics concepts because of their perceived inadequacy of knowledge, or due to lacking infrastructure in some schools. However, one of the most important components in making this process effective is teaching methods, activities, and materials. Yet another common challenge for both countries is the lack of technical infrastructure. Having not enough technology or out-of-date technologies makes it difficult to implement the informatics curriculum. Hence, inefficient digital competencies combined with insufficient technology resources, what could be achieved?

Therefore, based on the literature, content, and analysis of interview data following suggestions need to be considered for effective integration at the individual level:

- Efforts should support that teachers become aware of the fact that computational thinking and informatics concepts can be taught with or without the use of computers via various approaches;
- Strategies for overcoming resistance should be used to increase teachers' self-confidence and overcome fears of the use of technology;
- Professional development opportunities (practice-oriented) for integration into the educational process and for learning how to teach informatics should be continuously provided to classroom teachers;
- Coherent and systemic provision of teaching support (e.g. material/teacher book/practice by learning by doing) should be delivered.

Moreover, some suggestions at organizational level are also seem to be very important in the educational system:

- School principals and administrative staff should also have technological leadership competencies, so there should be acting as role models;
- Best practices should be shared together with the discussions of pros and cons.

Although this qualitative research study is limited to the volunteer teachers participating in the study, the efforts provided some different insights and revealed potential problems for many other countries. Hence, at least now it is known where to start and what possible obstacles could be faced in the future, so policy makers and other stakeholders can update agenda for effective integration of informatics in schools based on lessons learned from two countries.

Acknowledgement. The authors acknowledge the contribution of interviewed primary school teachers from both countries. This project has received funding from European Social Fund (project No 09.3.3-LMT-K-712-23-0083) under grant agreement with the Research Council of Lithuania (LMTLT).

References

1. Bocconi, S., et al.: Reviewing Computational Thinking in Compulsory Education (No. JRC128347). Joint Research Centre (Seville site) (2022)
2. Caskurlu, S., Yadav, A., Dunbar, K., Santo, R.: Professional development as a bridge between teacher competencies and computational thinking integration. In: Computational Thinking in Education, pp. 136–150. Routledge (2021)
3. Dagienė, V., Jevsikova, T., Stupurienė, G.: Introducing informatics in primary education: curriculum and teachers' perspectives. In: Pozdniakov, S.N., Dagienė, V. (eds.) ISSEP 2019. LNCS, vol. 11913, pp. 83–94. Springer, Cham (2019). https://doi.org/10.1007/978-3-030-33759-9_7
4. Dagienė, V., Jevsikova, T., Stupurienė, G., Juškevičienė, A.: Teaching computational thinking in primary schools: worldwide trends and teachers' attitudes. Comput. Sci. Inf. Syst. **19**(1), 1–24 (2021)
5. Denning, P.J., Tedre, M.: Computational thinking: a disciplinary perspective. Inf. Educ. **20**(3), 361–390 (2021)
6. Education development center: Lithuanian Informatics curriculum outline. https://informatika.ugdome.lt/lt/biblioteka/dokumentai/
7. Fagerlund, J., Häkkinen, P., Vesisenaho, M., Viiri, J.: Computational thinking in programming with Scratch in primary schools: a systematic review. Comput. Appl. Eng. Educ. **29**(1), 12–28 (2021)
8. Garzón-Artacho, E., Sola-Martínez, T., Romero-Rodríguez, J.M., Gómez-García, G.: Teachers' perceptions of digital competence at the lifelong learning stage. Heliyon **7**(7), e07513 (2021)
9. Gülbahar, Y., Kalelioğlu, F.: Bilişim Teknolojileri Ve Bilgisayar Bilimi: Öğretim Programı Güncelleme Süreci (ICT and computer science: Curriculum Improvement Process). Milli Eğitim Dergisi **47**(217), 5–23 (2018)
10. Gülbahar, Y., Ilkhan, M., Kilis, S., Arslan, O.: Informatics education in Turkey: national ICT curriculum and teacher training at elementary level. Commentarii informaticae didacticae: (CID) **6**, 77–87 (2013)
11. Howard, S.K., Tondeur, J., Ma, J., Yang, J.: What to teach? Strategies for developing digital competency in preservice teacher training. Comp. Edu. **165**, 104149 (2021)
12. Israel-Fishelson, R., Hershkovitz, A.: Persistence in a game-based learning environment: the case of elementary school students learning computational thinking. J. Educ. Comput. Res. **58**(5), 891–918 (2020)
13. Jocius, R., et al.: Code, connect, create: the 3C professional development model to support computational thinking infusion. In: Proceedings of the 51st ACM Technical Symposium on Computer Science Education, pp. 971–977 (2020)

14. Kert, S.B., Kalelioğlu, F., Gülbahar, Y.: A holistic approach for computer science education in secondary schools. Inf. Educ. **18**(1), 131–150 (2019)
15. Nordby, S.K., Bjerke, A.H., Mifsud, L.: Computational thinking in the primary mathematics classroom: a systematic review. Digit. Exp. Math. Educ. **8**, 1–23 (2022)
16. Tedre, M., et al.: Teaching machine learning in K-12 classroom: pedagogical and technological trajectories for artificial intelligence education. IEEE Access **9**, 110558–110572 (2021)
17. Zhang, L., Nouri, J., Rolandsson, L.: Progression of computational thinking skills in Swedish compulsory schools with block-based programming. In: Proceedings of the Twenty-Second Australasian Computing Education Conference, pp. 66–75 (2020)

Bebras Challenge in a Learning Analytics Enriched Environment: Hungarian and Indian Cases

Zsuzsa Pluhár[1]([✉]) [ID], Heidi Kaarto[2], Marika Parviainen[2], Sonia Garcha[3], Vipul Shah[4], Valentina Dagienė[5] [ID], and Mikko-Jussi Laakso[2] [ID]

[1] Eötvös Loránd University, Budapest, Hungary
`pluharzs@inf.elte.hu`
[2] University of Turku, Turku, Finland
`{heemkaa,mhparv,milaak}@utu.fi`
[3] CSpathshala, Pune, India
[4] Tata Consultancy Services, Pune, India
`v.shah@tcs.com`
[5] Vilnius University Institute of Educational Sciences, Vilnius, Lithuania
`valentina.dagiene@mif.vu.lt`

Abstract. Education needs to provoke young people to be active participants of modern society and contribute to changing and shaping the world. The international Bebras initiative, with over 70 countries participating, is one of the successful approaches involving school students in solving problems of computer science and deep thinking. In 2021, Finland, Hungary and India, supported by Lithuania, started a research study on solving Bebras tasks integrated into the Finnish virtual learning environment ViLLE using learning analytics. In this paper, we describe the methodology of the research study and two pilots conducted in Hungary and India with 1548 participants in total. A detailed analysis of Hungarian Bebras Challenge run in November 2021 in the ViLLE environment is provided. Results of 33,467 students aged 9–18 are discussed using task difficulty, gender, and time as the underlying variables. Also, a brief overview of feedback from teachers and students on using the ViLLE environment is given. The results from the pilots and from the Hungarian Bebras Challenge show that the ViLLE environment supports the task solving process of the Bebras Challenge and easy adaptive to different languages and task sets.

Keywords: Computational Thinking · Bebras Challenge · Computer science education · Task solving

1 Introduction

In recent years, Computational Thinking (CT) has been recognized as an essential skill for all citizens, as they are members of the digital era. Many researchers discuss the definitions of CT and the skills, components and main concepts included, adjust and develop new approaches, and establish learning content and assessment tools.

A. Bollin and G. Futschek (Eds.): ISSEP 2022, LNCS 13488, pp. 40–53, 2022.
https://doi.org/10.1007/978-3-031-15851-3_4

Four decades ago, Seymour Papert coined the term "computational thinking" and suggested that computers might enhance thinking and change patterns of knowledge accessibility [1]. Later Jeanette Wing [2] actively reformulated CT as thought processes and promoted these ideas in the design and analysis of problem solving.

Román-González et al. [3] distinguished three types of the definitions of CT: 1) generic definition that focus on CT as a thought process, 2) operational definition that describe what CT entails based on Selby and Woollard's five fundamental elements [4], and 3) curricular definitions derived from different frameworks.

According to various research works [1, 2, 5–7] CT is a way of thinking (thought process) for problem solving. However, researchers also specify that it is not just problem solving: the solution of the problem must be expressed in a way that allows a computational agent to carry it out [8–11].

A systematic review of empirical studies [9] highlighted that a lot of CT definitions are related to programming and computing. Numerous recent studies discussed that coding skills should be considered to be among basic skills and are of equal importance as reading and writing skills. Programming and CT are deeply intertwined: programming is used as means to learn the concepts and skills related to CT [10]. CT shifts the focus from simple coding to problem solving in various disciplines by emphasizing computational skills [11].

Thinking computationally means being able to approach and solve problems efficiently based on the principles and methods of computer science [12, 13]. Whichever view one takes of the definition of CT, it is important to be pragmatic regarding developing the best ways to teach it [14]. CT refers to an individual's ability to recognize aspects of real-world problems appropriate for computational formulation and evaluate and develop algorithmic solutions to those problems so that the solutions could be operationalized with a computer [15].

Most researchers [e.g., 16, 7, 17] use the 3-dimensional theoretical framework of CT components by Brennan and Resnick [18] based on CT Concepts, CT Practice, and CT Perspectives.

CT Concepts refer to the computational concepts, i.e., algorithm, pattern recognition, encryption, artificial intelligence, deadlock, decomposition, including programming terms, i.e., sequences, loops, parallelism, events, conditionals, operators, parameters, recursion, data structures [19–21].

During the last decade, CT Practices gained the most attention, although CT Concepts play a central role. Researchers tend to investigate different CT Concepts in their studies, so CT Concepts and Practices are tightly pinned together, the most cited are together.

Denning and Tedre [22] analyze CT from the perspective of computing's disciplinary ways of thinking and practicing and explain that CT is not a set of concepts for programming. It is a way of thinking that is honed through practice: the mental skills for designing computations to do jobs for us, and for explaining the world as a complex of information processes.

2   Bebras and ViLLE in Connection to Computational Thinking

2.1   The Bebras Challenge

The international Bebras Challenge promotes computer science or informatics and CT by solving short tasks based on CT concepts and is well-known in many countries as an informal school event [21]. Having organized the Bebras Challenge annually for almost twenty years, organizers (and authors of this paper) have noticed that students (and their teachers) consider the activities very exciting as they provide problem solving experience and insight into what lies beyond digital technology. The crucial point of the challenge is the tasks: they are based on the informatics concepts and help to understand what is beyond technology, they are short, attractive, and answerable in a few minutes.

The Bebras Challenge is aimed at all primary and secondary school students to promote informatics education and CT. Students solve 18 to 24 tasks in 45–55 min. There are different task sets for different age groups. Six age groups are used: Group I: Pre-Primary (grades 1–2), Group II: Primary or Little Beavers (grades 3–4), Group III: Benjamins (grades 5–6), Group IV: Cadets (grades 7–8), Group V: Juniors (grades 9–10) and Group VI: Seniors (grades 11–12). Participants are usually supervised by teachers who may integrate the Challenge into their teaching activities as well. More explanation of structure of the Bebras Challenge on www.bebras.org.

Solving short concept-based tasks makes informatics education more attractive for learners [23]. During the Challenge, students have the opportunity to test their skills among peers from different schools or even countries and make friends in a field that they have interest in. The challenge on CS and CT named "Bebras" (Lithuanian for "beaver") may be a key to the potential of informatics science knowledge and an attractive way to bind technology and education. A big challenge for Bebras is to organize easily accessible and highly motivating online problem-solving events in different countries.

The Bebras Challenge is mostly run online, however the Bebras model share most features of the CS unplugged approach [24, 25]: tasks can be solved also without the use of computer, learning by doing is the implied pedagogy, tasks cover the computing discipline in its broader sense.

2.2   ViLLE Learning Environment

ViLLE is a digital learning environment that has been developed at the Centre for Learning Analytics in the University of Turku, Finland [26]. ViLLE utilizes automatic assessment and offers immediate feedback for the students and comprehensive learning analytics for the teacher.

Technology enhanced learning with ViLLE is an effective way to improve students' mathematics skills both in short [27] and in long term [28] and this virtual environment is widely used in Finland, not only for math but for programming, languages, CT and more. The focus of developing the ViLLE environment and materials made by the Centre for Learning Analytics has always been in research. ViLLE supports over 150 automatically assessed exercise types which can be used by all ViLLE teachers. The Bebras Lodge, developed by Vilnius University team, is a tool to create interactive exercises/tasks [29], and from fall 2021 onwards it is also supported by ViLLE.

ViLLE has been used in national testing and therefore has a built-in Research System that can be flexibly adapted to different kinds of tests. This Research System supports anonymous accounts, all ViLLE exercise types, automatic assessment, time limits, monitoring students' performance and much more. The reason to use ViLLE in Bebras Challenge lies in these features.

The Bebras Challenge framework in ViLLE includes creating the tasks by the Bebras Community, transferring them into ViLLE, organizing them into courses for each age group, and connecting them into the Research System. With the Research System, we can create activation codes which the local administrators can send to the teachers. When the teachers use their activation codes, the Research System automatically creates a copy of the desired age group's course and the anonymous student accounts which the teacher can then easily give to the students. All the data from the students completing tasks is saved into the Research System from where it can be fully accessed. (See Appendix A).

3   Study Settings - The Research and Methodology

With the support from Lithuania, Finland, Hungary and India we have started research to study the future results of Bebras Challenge in many countries using the tools that ViLLE offers. We will analyze the differences in difficulty levels, types of tasks in CS, the influence of differences in cultural backgrounds, and the difficulty of the tasks in Bebras Challenges in multiple countries.

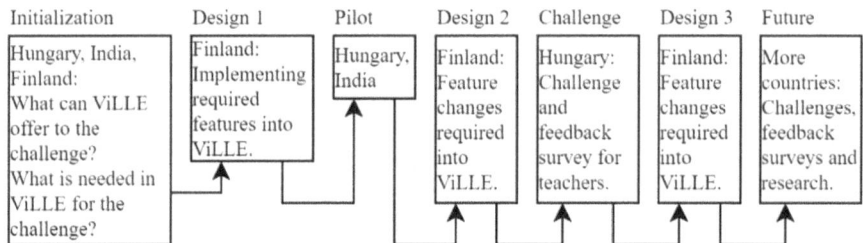

Fig. 1. The schema for the process of the research about the Bebras challenge in ViLLE

The study has been started with an initialization phase following a design phase, a pilot, a second design phase, an official Bebras Challenge and a third design phase.

The study focuses on the results of the pilots and first contest organized in this new environment for Bebras Challenge. The following researched question are raised:

- RQ1: How successfully could the ViLLE with Research System built-in support the student's task solving and teacher' organizational work in the Bebras Challenge?
- RQ2: Looking at the Bebras Challenge run in ViLLE, what patterns of students performance can be identified?

The pilots were used to test only the virtual environment ViLLE with research System built-in in order to see how students and teachers are comfortable using ViLLE environment for the Bebras Challenge. It is important to test human-machine interaction. After the pilots, based on feedback from the teachers we had a refine phase of the registration process and the documentation of ViLLE. We used questionnaires for feedback of the teacher's experiences with an overview of the student's reactions. Moreover, we studied the results (scores) and the time used to answer the questions in general separately for age groups with the emphasis on the genders.

The next challenges - will be organized with several countries in ViLLE - give the base of the research's lifecycle: after the closed challenges, we will analyze the results, the influence factor of the gender, the difficulty level of the tasks, and the cultural background of the participating countries. The results will support the improvement of the ViLLE system and the Bebras tasks development process.

4   Bebras Challenge in Hungary and India

The Hungarian Bebras Challenge [25, 30] has been organized since 2011 and the Indian since 2018. Besides promoting informatics and CT among students, the Bebras Challenges in Hungary and India seek to excite students about CT and interest them in exploring computing; help students develop problem solving skills and critical thinking skills; delink lack of computing infrastructure from teaching of computing by providing activities that can be done in classrooms and do not require computers; present ideas to teachers for school and after-school activities.

The number of participants is growing continuously from year to year in both countries. Students from both countries participate from many levels, ages 8–18 years and different kinds of schools. In each age group the participants have 45 min – based on the lesson lengths in the school systems – to solve tasks online. Students may participate in the Challenge at any time within a one or (since 2020) two weeks period ("Bebras week") in November.

India conducts the Challenge across 16 states and it is offered in English and also in the regional languages: Gujarati, Marathi, Kannada, Odia, Tamil and Telugu to enable inclusion: rural, semi-urban and government school students who are typically excluded and are given the opportunity to participate in the Challenge. The Hungarian Challenge uses only Hungarian as the official language of the country.

The Challenge in both countries uses tasks at three levels of difficulty (easy, medium, and hard) based on difficulty-sorting of International Bebras Task Workshop. In India, the tasks are multiple choice questions: five easy, five medium, and five hard level questions. Hungary also uses multiple choice questions, but four for the youngest and six for others in each difficulty level.

The answer options are mixed up, but the tasks are always in the same order according to their difficulty. Participants can switch between tasks while they compete, but they get the easiest tasks first and the hardest tasks last. ViLLE is designed so that the students are encouraged to solve the tasks from the first to the last and offers an option choose the next one. This also means that most of the students answer all of the questions.

As the Bebras Challenge has negative marks for incorrect answers, the students begin the Challenge with bonus marks so that no student feels discouraged by getting negative marks in the end.

5   The Pilots in Hungary and India

A pilot for the ViLLE environment for Bebras practice Challenge in Indian and Hungarian schools was conducted in October 2021. Owing to the pandemic the students in a majority of the states in India were not physically back in school, so the selection of Indian schools was based on students having access to digital technology and organized in a hybrid mode (online and in-person). The Hungarian pilot was organized in schools in-person.

5.1   Process of the Piloting

Three Indian schools had volunteered for the pilot and registered the students but in the end, students from only two schools could participate in the pilot. Owing to the constraints due to floods in Tamil Nadu in October/November which led to complete school closure, the students from the third school were unable to participate. The students from the selected schools had not participated in the Bebras India Challenge earlier and they were exposed to the Bebras tasks for the first time. Hungarian schools were selected randomly from schools participating in previous years.

India had selected twenty tasks for the pilot, and it was conducted across two groups: Group II (grades 3 and 4) and Group III (grades 5 and 6). They used ten common tasks across the two groups with 15 tasks in each group's Challenge. The Bebras tasks were multiple choice questions: five easy, five medium, and five hard level questions. Hungary didn't determine the age of the participants in the pilot but used only the tasks from last year in the Benjamin (age 11–12) group. In both countries the duration of the Pilot was 45 min.

The Indian team organized an online orientation session for the teachers of the participating schools with a walk through of the process for conducting this pilot in the ViLLE environment. The activation codes were shared with the teachers to set up the competition for their class and to manage the competition afterwards. Since the number of students per class is high in India, the activation code was the same for a maximum of 75 students taking the same Bebras Challenge groups (II or III). The registration of the students was managed by the schools.

Hungary translated and prepared documents and sent them to the teachers with the activation codes. The registration of the students was organized by the teachers. The number of students were between ten and twenty in each group because the teachers took the pilot in the school, during a lesson.

5.2   Results of the Pilots

The feedback collected for pilots in both Hungary and India were qualitative. The feedback was used to design a feedback form for the actual Challenge that provided more quantitative feedback.

The Pilot in India.
A total of 196 students from two schools participated in the Indian Challenge with an average participation of 98 students per school. There were 107 students (52 girls and 55 boys) in the Group II and 89 students (43 girls and 46 boys) in the Group III from the

two schools. The average score in Group II was 66.034 (girls 66.0577, boys 66.0182) and the highest score was 149 and the lowest score was 19 of a maximum score of 165. In Group III the average score was 61.0674 (girls 55.8837, boys 65.913) and the highest score was 165 and the lowest score was 6 of a maximum score of 165.

The teachers had informed the students that this would be like a game that they would be playing and showing their skills while solving playful tasks. The students liked the ViLLE environment, had no difficulty using it, and they enjoyed solving the tasks as this was their first time. The pilot familiarized the students with the Bebras tasks, and the virtual environment.

During the feedback, the teachers shared that the registration process on the ViLLE environment was fairly simple. The main problem was that in the hybrid mode (in-person and virtual teaching) it became difficult to monitor and motivate students to take participation. The teachers also shared that handing over the student login slips to the boys and girls separately was another work that teachers felt was hectic. Another problem the teachers faced was with schools reopening after nearly a year and a half, the labs have not been functional and internet connectivity was a problem too.

The Pilot in Hungary.
There were 1351 students (708 girls and 641 boys) from 40 schools participating in the Hungarian pilot. On average, 33.775 students participated from each school, and the maximum number of students from one school was 83.

The average score was 115.3412 (girls 114.3588, boys 116.3701). The highest average score per school was 153.343 and the lowest 82.692. The students were coming from several age groups and most of them probably knew the tasks from last year. Therefore, the emphasis of the Hungarian pilot was on the new environment and preparing to use it for the main Challenge.

At the end of the pilot the teacher's feedback was collected by a questionnaire and analyzed. The registration process and the translations, user documentation were improved based on the feedback.

6   Results in Hungarian Challenge

In 2021 Hungary used the ViLLE environment for the official Bebras Challenge. That included the translation of the ViLLE environment and integration of the tasks in Hungarian. The Bebras Challenge was prepared as in previous years: the tasks were not interactive; the students could choose one answer from four options. The order of the tasks was the same (easy, medium and hard) and the order of the answer options were random.

Over 30 thousand participants (N = 33467, from 305 schools) across the whole country and Hungarian language schools joined from outside Hungary as well. More schools attend from the capital and western regions (as shown in the map – see Appendix B).

The largest increase in the number of participants can be found in the Little Beavers (age 9–10) age group with a 36% increase compared to the year 2020. The largest number of participants (12491 students) was in the Junior (age 15–16) age group. (See Appendix B).

This time the youngest had 12 tasks (four in each difficulty level) and they could reach a total of 144 points. The other four age groups had 18 tasks as usual (six in each difficulty level) and they could reach a total of 216 points. Figure 1 shows the number of students in each group of total scores by age group (except Little Beavers because of lower total score) (Fig. 2).

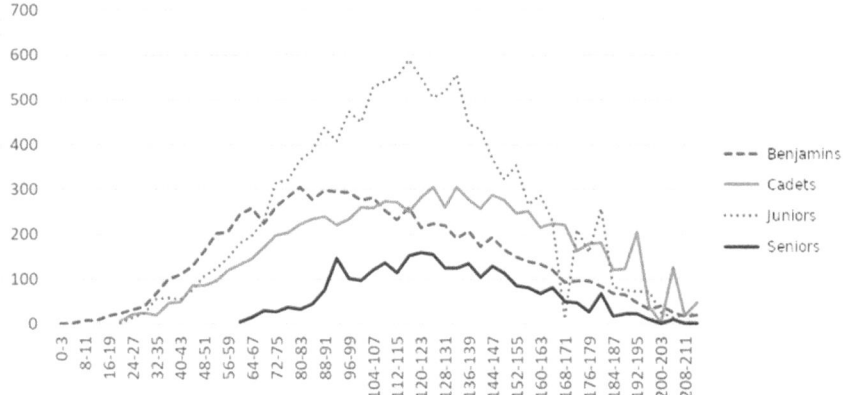

Fig. 2. The number of students per age group achieving a given scores.

More analyses can be done on the standard deviation of scores in each age group (see Appendix B). There is no significant difference between the scores of girls (f) and boys (m) in the two youngest age groups (Benjamins: $N_m = 4104$; $N_f = 3859$; $Z = -.553$; $p = .580$ and Little Beavers: $N_m = 927$; $N_f = 833$; $Z = -1.492$; $p = .136$.). The total time working with tasks is significantly higher in the case of girls (Benjamins: $N_m = 4104$; $N_f = 3859$; $Z = -7.413$; $p < .001$ and Little Beavers: $N_m = 927$; $N_f = 833$; $Z = -3.645$; $p < .001$).

The three oldest groups' scores show significant differences between boys and girls. (Cadets: $N_m = 4194$; $N_f = 4274$; $Z = -4.027$; $p < .001$; Juniors: $N_m = 6562$; $N_f = 5929$; $Z = -2.801$; $p = .005$ and Seniors: $N_m = 1976$; $N_f = 808$; $Z = -5.178$; $p < .001$). There is no significant difference between the total times of working with tasks of the boys and girls.

In recent years there has not been significant differences with the genders, so we analyze deeper what could be the influencer this year.

The average time "in task" (Little Beavers: 823.80; Benjamins: 1616.83; Cadets: 1528.83; Juniors: 1801.53; Seniors: 1943.78) shows the increase of task's difficulty.

An analysis of variance (ANOVA) on these scores and the times in the tasks yielded significant variation among age groups ($F_{score} = 811.159$, $F_{time} = 1640.349$, $p < .001$). A post hoc Tukey HSD test ($\alpha < .05$) showed that both the main scores ([Seniors] < [Benjamins] < [Juniors] < [Cadets]) and the used times ([Cadets] < [Benjamins] < [Juniors] < [Seniors]) for age groups differed significantly at $p < .001$. The students in the oldest age group had significantly smaller scores (in mean) and they spent the longest time with tasks. The students in the Cadets spent significantly less time and had a significantly higher mean score. The number of students with the maximum score in Cadets predicted this result. Our future analysis about the difficulty level and hardness of the tasks may explain these results.

6.1 Teachers' Feedbacks

Teachers were asked to fill a questionnaire about their experiences using the ViLLE environment. The questions were mostly open ended or scaled by a five-point Likert scale. About 20% of the teachers (N = 77) answered the questionnaire from a third of the schools.

96% of the teachers have found the speed of ViLLE environment is very fast or fast. Comparing to previous years (using the Hungarian platform) percentage increases more than 20%. The 77 teachers coordinated 645 groups in total (approximately 9600 students based on the average number of students in a group). The highest number of groups registered by one teacher was 36, the average 8.38 and the median 6. 26 (33.77%) teachers had more than 10 groups. The registration process for teachers with 1–5 groups was easier than in previous competitions. The teachers coordinating more than 10 groups found the process more complicated and longer because of separated work processes: they had an account for each group separately.

The students found that the ViLLE environment is easier to use and more comfortable and safer. It means if an error (i.e., restarting the browser or the computer, power outage) occurred, they could continue the work later without losing the answers. They enjoyed the friendly environment with "smiley" in the feedback.

7 Discussion and Future Work

The feedback and the successful running of the Hungarian Challenge shows that the ViLLE environment (with Research System built-in) provides a perfect basis to organize the Bebras Challenge. The users from age 9 to 18 years old used the system without problem and the teacher's administration work went smoothly.

From the participated five age groups the three oldest groups' (IV, V, and VI) scores show significant differences between boys and girls. It was not surprising that the analysis revealed that boys outperform girls in the Hungarian Bebras Challenge. Hubwieser et al. [31] noticed the same in the German Bebras Challenge of 2014. Interestingly when Budinska et al. [32] studied the Slovakian Bebras Challenges of 2012 to 2017 (age group 7–10), they found out that girls usually perform better in easier tasks, while boys excel in more difficult tasks. Overall, however, in this age group, the girls achieved higher scores on average. Both research teams were able to identify features that made a task more suitable to girls.

The presented analysis of the results (scores) and used times suggest studying the tasks deeper in view of the age groups and difficulty levels. The second research question (RQ2) was answered partly, so we are going to finish the third part of the analysis written in the "Study settings" chapter.

Some improvements can be useful to help teachers with more groups based on the feedback and integrating the interactivity could open a new horizon for the students. In the future, we are planning on having more countries participating in Bebras Challenge in ViLLE and conducting international research based on the results. Additionally different types of interactive tasks are going to be used. Next steps of our research can be the study and analysis of the differences in difficulty levels, the variety of categories of the type, the cultural background influence and the hardness of a Bebras tasks in multiple countries.

Appendix A - Schemas

The schema for the process of organizing a Bebras Challenge in ViLLE

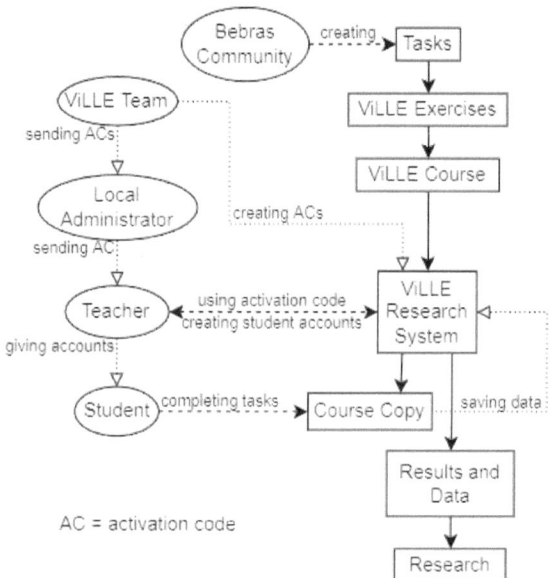

The process of organizing a Bebras Challenge in ViLLE.

Appendix B - Figures Representing Data in Hungarian Bebras Challenge 2021

Map of the participation in the Bebras Challenge 2021 in Hungary

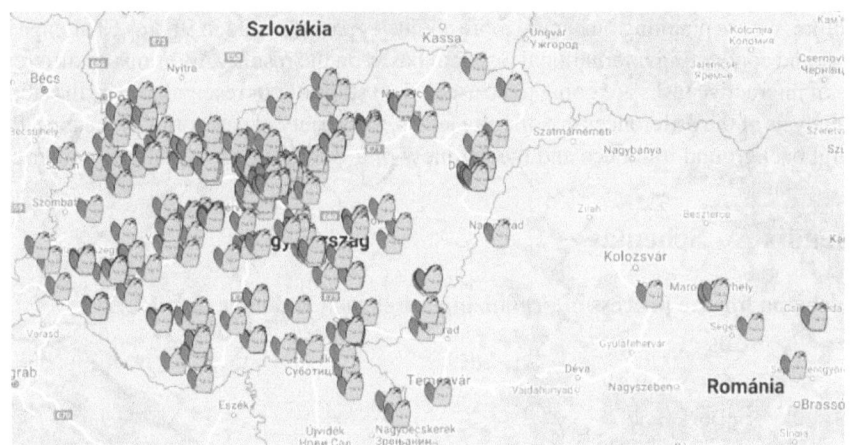

The map of schools participating in the Hungarian Bebras Challenge in 2021. The beavers mark the participating schools.

Participation in Hungarian Bebras Challenge by age group (2011–2021).

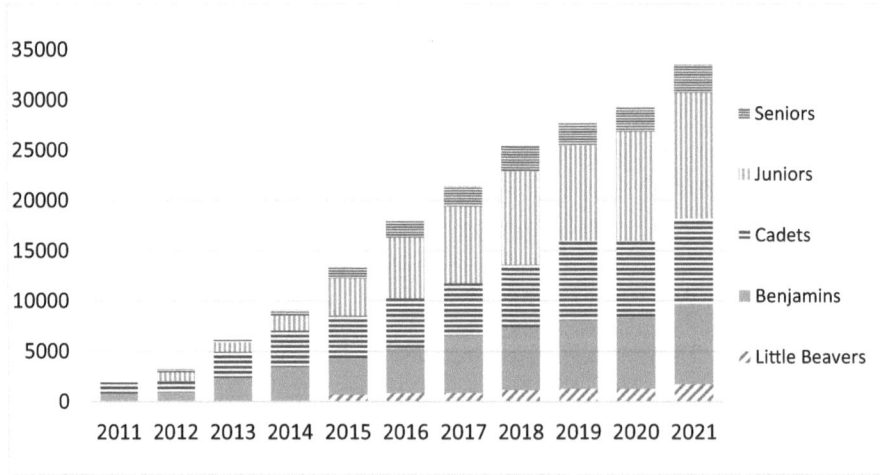

The number of participants in each age group in the Hungarian Bebras Challenge from 2011 to 2021.

The standard deviations of scores in each age group

Age group	Number of participants (girls/boys)	Mean of scores (girls/boys)	Standard deviation of scores (girls/boys)
Little Beavers (9–10)	1760 (833/927)	66.60 (67.61/65.69)	27.11 (26.64/27.50)
Benjamins (11–12)	7963 (3859/4104)	104.31 (104.41/104.22)	41.56 (40.56/42.47)
Cadets (13–14)	8469 (4275/4194)	118.35 (118.30/118.40)	42.09 (40.66/43.51)
Juniors (15–16)	12491 (5929/6562)	111.25 (110.21/112.20)	35.62 (34.61/36.49)
Seniors (17–18)	2784 (808/1976)	93.98 (89.06/96.00)	30.80 (28.19/31.59)

References

1. Papert, S.: Mindstorms: Children, Computers, and Powerful Ideas. Basic Books Inc., New York (1980)
2. Wing, J.: Computational thinking. Commun. ACM **49**, 33–35 (2006)
3. Román-González, M., Pérez-González, J.-C., Jiménez-Fernández, C.: Which cognitive abilities underlie computational thinking? Criterion validity of the Computational Thinking Test. Comput. Hum. Behav. **72**, 678–691 (2017). https://doi.org/10.1016/j.chb.2016.08.047
4. Selby, C., Woollard, J.: Computational thinking: the developing definition (2013). https://eprints.soton.ac.uk/356481
5. Grover, S., Pea, R.: Computational thinking: a competency whose time has come. In: Sentance, S., Barendsen, E., Carsten, S. (eds.) Computer Science Education: Perspectives on Teaching and Learning in School, pp. 19–38. Bloomsbury, London (2018)
6. Hazzan, O., Ragonis, N., Lapidot, T., Rosenberg-Kima, R.: Computational thinking. In: Guide to Teaching Computer Science, pp. 57–74. Springer, Cham (2020). https://doi.org/10.1007/978-3-030-39360-1_4
7. Zhang, L., Nouri, J.: A systematic review of learning computational thinking through Scratch in K-9. Comput. Educ. **141**, 1–25 (2019). https://doi.org/10.1016/j.compedu.2019.103607
8. Corradini, I., Lodi, M., Nardelli, E.: Conceptions and misconceptions about computational thinking among Italian primary school teachers. In: ICER – Proceedings of ACM Conference International Computing Education Research, pp. 136–144 (2017). https://doi.org/10.1145/3105726.3106194
9. Tang, X., Yin, Y., Lin, Q., Hadad, R., Zhai, X.: Assessing computational thinking: a systematic review of empirical studies. Comput. Educ. **148**, 1–22 (2020). https://doi.org/10.1016/j.compedu.2019.103798
10. Metcalf, S.J., et al.: Assessing computational thinking through the lenses of functionality and computational fluency. Comput. Sci. Educ. **31**(2), 199–223 (2021). https://doi.org/10.1080/08993408.2020.1866932
11. Basu, S., Biswas, G., Kinnebrew, J.S.: Learner modeling for adaptive scaffolding in a computational thinking-based science learning environment. User Model. User-Adap. Inter. **27**(1), 5–53 (2017). https://doi.org/10.1007/s11257-017-9187-0

12. Arfé, B., Vardanega, T., Ronconi, L.: The effects of coding on children's planning and inhibition skills. Comput. Educ. **148**, 1–16 (2020). https://doi.org/10.1016/j.compedu.2020.103807

13. Palts, T., Pedaste, M.: A model for developing computational thinking skills. Inform. Educ. **19**, 113–128 (2020). https://doi.org/10.15388/INFEDU.2020.06

14. Curzon, P., Bell, T., Waite, J., Dorling, M.: Computational thinking. In: Robins, A.V., Fincher, S.A. (eds.) The Cambridge Handbook of Computing Education Research, pp. 513–546. Cambridge Univ. Press, Cambridge (2019). https://doi.org/10.1017/9781108654555.018

15. Eickelmann, B., Labusch, A., Vennemann, M.: Computational thinking and problem-solving in the context of IEA-ICILS 2018. In: Passey, D., Bottino, R., Lewin, C., Sanchez, E. (eds.) OCCE 2018. IAICT, vol. 524, pp. 14–23. Springer, Cham (2019). https://doi.org/10.1007/978-3-030-23513-0_2

16. Rich, P.J., Mason, S.L., O'Leary, J.: Measuring the effect of continuous professional development on elementary teachers' self-efficacy to teach coding and computational thinking. Comput. Educ. **168**, 104196 (2021). https://doi.org/10.1016/j.compedu.2021.104196

17. Zhang, L., Nouri, J., Rolandsson, L.: Progression of computational thinking skills in Swedish compulsory schools with block-based programming. In: Proceedings of the Twenty-Second Australasian Computing Education Conference, pp. 66–75 (2020). https://doi.org/10.1145/3373165.3373173

18. Brennan, K., Resnick, M.: New frameworks for studying and assessing the development of computational thinking. In: Proceedings of the 2012 Annual Meeting of the American Educational Research Association, vol. 1, pp. 1–25, Vancouver (2012)

19. Hromkovič, J., Lacher, R.: The computer science way of thinking in human history and consequences for the design of computer science curricula. In: Dagiene, V., Hellas, A. (eds.) ISSEP 2017. LNCS, vol. 10696, pp. 3–11. Springer, Cham (2017). https://doi.org/10.1007/978-3-319-71483-7_1

20. Hromkovič, J., Kohn, T., Komm, D., Serafini, G.: Combining the power of Python with the simplicity of logo for a sustainable computer science education. In: Brodnik, A., Tort, F. (eds.) ISSEP 2016. LNCS, vol. 9973, pp. 155–166. Springer, Cham (2016). https://doi.org/10.1007/978-3-319-46747-4_13

21. Dagienė, V., Stupurienė, G.: Bebras - a sustainable community building model for the concept based learning of informatics and computational thinking. Inf. Educ. **15**(1), 25–44 (2016)

22. Denning, P.J., Tedre, M.: Inf. Educ. **20**(1), 361–390 (2021). https://doi.org/10.15388/infedu.2021.21

23. Dagiene, V., Futschek, G., Stupuriene, G.: Creativity in solving short tasks for learning computational thinking. Constructivist Found. **14**(3), 382–396 (2019)

24. Bell, T., Vahrenhold, J.: CS unplugged – how it is used, and does it work?. In: Böckenhouer, H.-J., Komm, D., Unger, W. (eds.) Adventures Between Lower Bounds and Higher Attitudes: Essays Dedicated to Juraj Hromkovič on the Occasion of His 60th Birthday, pp. 497–521. Springer, Cham (2018). https://doi.org/10.1007/978-3-319-98355-4_29

25. Pluhár, Zs.: Extending computational thinking activities. Olympiads Inf. **15**, 83–89 (2021)

26. Laakso, M.-J., Kaila, E., Rajala, T.: ViLLE – collaborative education tool: designing and utilizing an exercise-based learning environment. Educ. Inf. Technol. **23**(4), 1655–1676 (2018). https://doi.org/10.1007/s10639-017-9659-1

27. Kurvinen, E., Dagienė, V., Laakso, M.-J.: The impact and effectiveness of technology enhanced mathematics learning. In: Dagienė, V., Jasutė, E. (eds.) Constructionism 2018: Constructionism, Computational Thinking and Educational Innovation: Conference Proceedings. Vilnius University, pp. 351–363 (2018)

28. Kurvinen, E., Kaila, E., Laakso, M., Salakoski, T.: Long term effects on technology enhanced learning: the use of weekly digital lessons in mathematics. Inf. Educ. **19**(1), 51–75 (2020). https://doi.org/10.15388/infedu.2020.04

29. Dagienė, V., Stupurienė, G., Vinikienė, L.: Implementation of dynamic tasks on informatics and computational thinking. Baltic J. Mod. Comput. **5**(3), 306–316 (2017)
30. Pluhár, Zs., Gellér, B.: International informatic challenge in hungary. In: Auer, M.E., Guralnick, D., Simonics, I. (eds.) Teaching and Learning in a Digital World: Proceedings of the 20th International Conference on Interactive Collaborative Learning, pp 425-435. Springer, Cham (2017). https://doi.org/10.1007/978-3-319-73204-6_47
31. Hubwieser, P., Hubwieser, E., Graswald, D.: How to attract the girls: gender-specific performance and motivation in the bebras challenge. In: Brodnik, A., Tort, F. (eds.) ISSEP 2016. LNCS, vol. 9973, pp. 40–52. Springer, Cham (2016). https://doi.org/10.1007/978-3-319-467 47-4_4
32. Budinská, L., Mayerová, K., Veselovská, M.: Bebras task analysis in category little beavers in slovakia. In: Dagiene, V., Hellas, A. (eds.) ISSEP 2017. LNCS, vol. 10696, pp. 91–101. Springer, Cham (2017). https://doi.org/10.1007/978-3-319-71483-7_8

How is Two Better Than One?
An Observational Study on the Impact of Working in Pairs When Solving Bebras Tasks

Carlo Bellettini[ID], Violetta Lonati[ID], Mattia Monga[(✉)][ID],
and Anna Morpurgo[ID]

Università degli Studi di Milano, Milan, Italy
{bellettini,lonati,monga,morpurgo}@di.unimi.it

Abstract. Every year the Bebras challenge proposes small tasks to students, based on CS concepts. In Italy, in 2021 for the first time, it was possible to choose whether to participate in the challenge individually or in teams of two students. The team size was expected to affect the performance of students; in particular working in pairs was expected to increase the probability of solving the tasks correctly. We carried out an observational study on the results of the 2021 Bebras challenge in Italy, aiming at investigating and measuring the effects of team size on the performance. The findings confirm that working in pairs generally improves the team performance, but the impact is much smaller than expected. We observed that the positive effect of collaboration is greater with younger pupils and somewhat decreases when age increases. We identified and discussed the features of tasks where the impact was more relevant, and where this trend was more evident. We also propose some hypotheses, to analyze in future qualitative studies, to interpret the results.

Keywords: K12 · Bebras challenge · Observational studies

1 Introduction

In cognitive theory, many studies suggest that collaboration between peers enhances learning. This seems particularly true in STEM (Science, Technology, Engineering, and Math, including Computer Science[1]) education, where several popular methodologies, *e.g.,* collaborative and problem-based learning, are in fact based on this assumption [7]. Moreover, *pair programming* is a common practice in the so called *agile* approaches to software engineering and is often also adopted in many educational contexts [9]. Faced with a problem, and working in a small group to solve it, pupils can explore the problem and its features, and thus devise, analyse and contrast solving strategies, in a process of

---

[1] See for example https://www.ed.gov/stem.

A. Bollin and G. Futschek (Eds.): ISSEP 2022, LNCS 13488, pp. 54–65, 2022.
https://doi.org/10.1007/978-3-031-15851-3_5

collaborative knowledge building. Collaborative learning is also said to increase motivation and engagement [4].

For these reasons, since its first edition in our country, the participation to the Bebras Challenge was organized around teams. The Bebras International Challenge on Informatics and Computational Thinking[2] is a yearly contest organized in several countries since 2004 [1,3], with almost three million participants worldwide. The contest, open to pupils of all school levels (from primary up to upper secondary), is based on tasks rooted on core informatics concepts, yet independent of specific previous knowledge such as for instance that acquired during curricular activities. According to the (informal) feedback we received from many teachers, the Bebras challenge is able to engage pupils — even those who show less motivation in usual school activities — and to often activate lively discussions and interesting exchanges within the groups.

In 2021, due to the pandemic and to adhere to the social distancing rules, the participation in teams could have been problematic. Hence, we allowed participation in "teams" of individuals (*"singles"*) or teams of pairs (*"doubles"*), to meet different schools' needs and organizational constraints. We analyzed the results of overall 19'490 teams, 11'055 singles, 8'435 doubles, who participated over 5 different age categories (for a total of 27'925 students). All teams in the same age category were asked to solve the same suite of tasks, regardless of their team size, but their results were ranked distinctly. Together with the submitted answers, the Bebras platform [12] we use collects data concerning the interactions of teams with the platform itself (how much time pupils spend on each specific task, whether and when they go back and review/change their answer to an already completed task, whether they perform actions that generate feedback from the system, and so on). This offered us the chance to conduct an observational study about the effects of team size on the performance. Our main research question is:

RQ - How does the team size affect the performance of Bebras solvers?

Our initial hypothesis was that teams formed by two pupils would perform *better* than the individuals. The research question can then be articulated in two further sub-questions:

RQ1 - For which categories of pupils does working in pairs have the most positive impact?

RQ2 - For which kinds of tasks does working in pairs have the most positive impact?

Our findings confirm the initial hypothesis, but show that the effect of team size on performance is in general less than expected. Moreover we observe that such effect occurs differently according to the age of pupils and the features of tasks. We discuss these differences and state some hypotheses that may explain them. Such hypotheses are to be explored further in a future in-depth qualitative study.

---

[2] See http://bebras.org.

The paper is organized as follows. In Sect. 2 we present the collected data and the methods we used to analyze them. In Sect. 3 we present our findings: in Sect. 3.1 we compare the performances of singles versus doubles, and analyze the role of age categories on the differences between such performances; in Sect. 3.2 we show which tasks benefit most from collaboration and detect relevant features of these tasks. In Sect. 4 we acknowledge the limitations of our study. After discussing some related works in Sect. 5, conclusions are drawn in Sect. 6.

2   Methodology

This is an *observational study*. This means that the data we analyzed were not purposely gathered with a designed experiment, instead they were collected during the Bebras challenge held in November 2021. We first describe the data set and then present the methods we used for the analysis.

2.1   Dataset

The data were collected in order to manage the participation of schools, administer the contest, monitor and possibly fix issues arising during the challenge (*e.g.,* malfunctioning, cheating, loss of data), and to perform statistical analyses.

Schools participating in the challenge were informed that students' data were collected and they consented to their use for research and statistical presentation of the results. In fact no national ranking is ever published, only aggregated data (but the teachers can see the performances of all the teams of their school and the ranking within an institution). All analyzed data were anonymized by deleting most of the personal identifying data: we only retained the regional provenance of teams in order to analyze their geographical distribution (we cover all the administrative regions of our school system).

The dataset contains information about the performance of each team in the contest. Each team belongs to one category (among five) according to their components' age. Teams can have different sizes, *i.e.,* there are teams formed by a pair of students ("doubles") or just a single individual ("singles"). The numbers of teams considered in our analysis are reported in Table 1, grouped by category and team size. All teams in the same age category were asked to solve the same suite of 12 tasks, independently of the team's size. Some tasks appeared in more than one category. For each team, we know for which of the tasks assigned to their category they answered correctly and for which not. Table 1 also presents the average ratio of correct answers to tasks in each category. Finally, we have data concerning how the teams interacted with the contest platform while solving the task; the kind of data we can collect are described in [12]. All the anonymous data we analyzed are available at https://doi.org/10. 13130/RD_UNIMI/WT9NHU for independent studies and cross-validation.

Table 1. Number of participants and success ratio (average over all tasks) for each category and team size. The last column reports the increment in the average success ratio obtained with doubles compared to singles.

Category	N. of teams	Success ratio	Δ doubles − singles
IV–V grade	2740	45.4%	8.6%
Singles	1092	40.2%	
Doubles	1648	48.8%	
VI–VII grade	7544	35.1%	5.7%
Singles	4545	32.8%	
Doubles	2999	38.5%	
VIII grade	3431	32.6%	2.6%
Singles	2031	31.5%	
Doubles	1400	34.1%	
IX–X grade	3450	32.7%	1.7%
Singles	2245	32.1%	
Doubles	1205	33.8%	
XI–XIII grade	2325	39.3%	4%
Singles	1142	37.3%	
Doubles	1183	41.3%	

2.2 Analysis Methods

We considered each task solution as a random event with a binary outcome: solved or not solved. To simplify the problem, we considered each task as an independent event. In order to estimate the probability of answering correctly, we used a Markov chain Monte Carlo approach, then we relied on this estimation to compare the performances of singles and doubles, and to contrast them with relevant combinatorial benchmarks.

Estimating the probability of a correct solution. Let us consider the probability of the event "correctly solving any Bebras task", that is the probability that covariate C (as for "correctness") is 1 (C is 0 if the team gives the wrong answer to the task). We can estimate such probability by sampling a probabilistic model in which C is a random variable with a Bernoulli likelihood with an unknown parameter p, the probability of solving any task (not a specific one); then we estimated the *a posteriori* (*i.e.*, having seen the actual data) distribution of p with a Markov chain Monte Carlo approach (to implement our model we used the probabilistic programming language Stan[3]). We used a uniform prior for p: this assumption is rough (p is certainly different from 0 and 1, for example), but it matters very little in the process since we have a lot of data and the estimation of the posterior distribution is in fact rather robust w.r.t. to the choice of the prior. One could estimate p by simply taking the average success ratio (see Table 1), but this is a *point estimation* with no information about the uncertainty of its value[4]. The method we followed [5], instead, gives the whole

---

[3] See https://mc-stan.org/.
[4] One can estimate also the variance of p in order to have a measure of the variability, but an estimation of the error with respect to the "true value" needs inevitably some assumption on the underlying distribution.

distribution of p that we can use to estimate uncertainty intervals (for example the range in which 99% of the distribution lies), valid under the explicit model we used (*i.e.*, C is Bernoulli distributed with unknown p).

In particular we used this method to estimate the distribution of probability $p_{singles}$ of the event "correctly solving any Bebras task" for any singles, and of probability $p_{doubles}$ of the event "correctly solving any Bebras task" for any doubles. More formally,

$$p_{singles} = p(C = 1 \mid teamsize = 1)$$
$$p_{doubles} = p(C = 1 \mid teamsize = 2)$$

Analysis of the impact of team size on the correctness of answers. We expect that working in pairs improves the performance of teams. More formally, we expect $p_{singles} < p_{doubles}$. From a purely combinatorial viewpoint, we can say that the collaboration in a pair is *fully successful* if the pair is able to answer correctly whenever there is at least one of its members that would answer correctly alone. This means that the pair is able to recognize the correct answer even when the other member, alone, would answer incorrectly. We say that the collaboration is *fully harmful* in the opposite, worst-case, scenario, that is if the pair gives a wrong answer except when both pupils are able to answer correctly alone. This means that when only one of the pupils, alone, were able to answer correctly, the pair would not be able to recognize the correct answer and that the wrong answer always prevails. In general, we expect that the collaboration takes place at an intermediate level between fully harmful and fully successful. In probability terms, a pair is right with probability $p_{worst} = p_{singles}^2$ if the collaboration is fully harmful, and with probability $p_{best} = 1 - (1 - p_{singles})^2 = 2p_{singles} - p_{singles}^2$ if it is fully successful. We will compare these combinatorial benchmarks with the actual performance of doubles.

Teams with unusual team size. The size of teams is decided by teachers. Organizational issues (*e.g.*, the availability of a sufficient number of computers) probably had a relevant role in this choice. Moreover, constraints on the size of teams were due to the pandemic special regulations, which varied among regions and school levels (*e.g.*, remote attendance was avoided in primary school, whereas hybrid attendance was very common in high school); during the contest, some classes were attending in person, others remotely, and others used hybrid attendance. Besides these external factors, teachers were free to choose between singles and doubles. For instance they may have built teams randomly, or may have let their students choose how and with whom to participate, but they may also have considered students' prior ability to form balanced teams; in particular, they may have decided to pair students with special educational needs with a mate, or to let excellent students compete alone in a single team. We do not have any direct information about the criteria each teacher used to compose their teams. However, we know the number of double and single teams for each teacher, and this allows us to distinguish the cases where the choice of a different size is dictated by situations like the class having an odd number of pupils from special cases where the composition of a team turns out to be unusual for that teacher, and hence might be related to the ability of its components. We focus on the set of teams that have a typical composition among those of the same teacher: these are the singles of teachers who have at least 75% of singles and the doubles of teachers who have at least 75% of doubles. For these teams ("typical teams") we have reasons to believe the composition type is not biased by the members' prior ability, whereas the others could have been formed according to some specific ability-related criterium. In fact, we found 17'871 teams with a typical composition type, and only a small proportion of all teams (8%) with an untypical composition type. It is still possible that some criterium to compose teams was adopted at the school level, but we believe this is improbable in the general case, since mixing people from different classes is normally quite difficult in our school system and unlikely for a non competitive contest like Bebras.

The role of content and task features. The difference in performances between singles and doubles varies from task to task, and we identified the tasks where the impact of team size on correctness was higher. We analyzed the specific content and features of those tasks and formulated some hypotheses that would explain the higher impact for those tasks. We provided some support for these hypotheses by analyzing the data concerning the interaction of teams with the contest platform when solving those tasks.

3 Findings

3.1 Comparing Performances of Singles and Doubles

Figure 1 shows the distribution of probabilities $p_{singles}$ and $p_{doubles}$ together with the combinatorial benchmarks corresponding to fully successful and fully harmful collaborations.

The probability of solving a task (any task) for singles is on average 33%. Our model estimated that 99% of the probability mass (High Density Interval, HDI) lies between 0.33 and 0.34. Doubles have a higher probability (the mean of $p_{doubles}$ is 39%, HDI: 0.39–0.40) and the difference is on average +6% (HDI: 0.054–0.065). However, 39% is much *lower* than 56%, the value one would have with fully successful collaborations ($p_{best} = 1 - (1 - p_{singles})^2$).

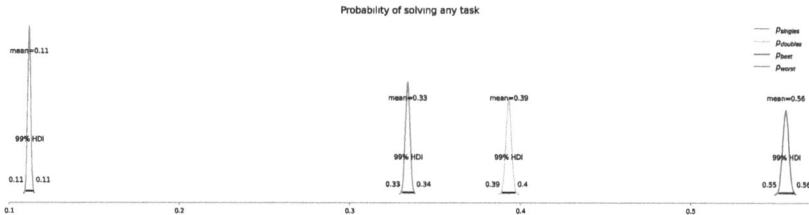

Fig. 1. The distribution of $p_{singles}$ (blue) is less than the distribution of $p_{doubles}$ (orange). The figure shows also the benchmarks for fully harmful collaboration (red) and fully successful collaboration (green). (Color figure online)

Notice that the diagram represents distributions of probability for p_{worst}, $p_{singles}$, $p_{doubles}$, p_{best}, from left to right. According to our model, the estimated values (their distributions) of $p_{singles}$ and $p_{doubles}$ do not overlap (in particular their HDI do not overlap), thus the difference (and its measure) is supported by a clear evidence, if our statistical model is a sensible abstraction of our domain.

In summary, it is clear that the overall effect of collaboration is positive, however it is limited with respect to the combinatorial benchmark p_{best} of the fully successful collaboration.

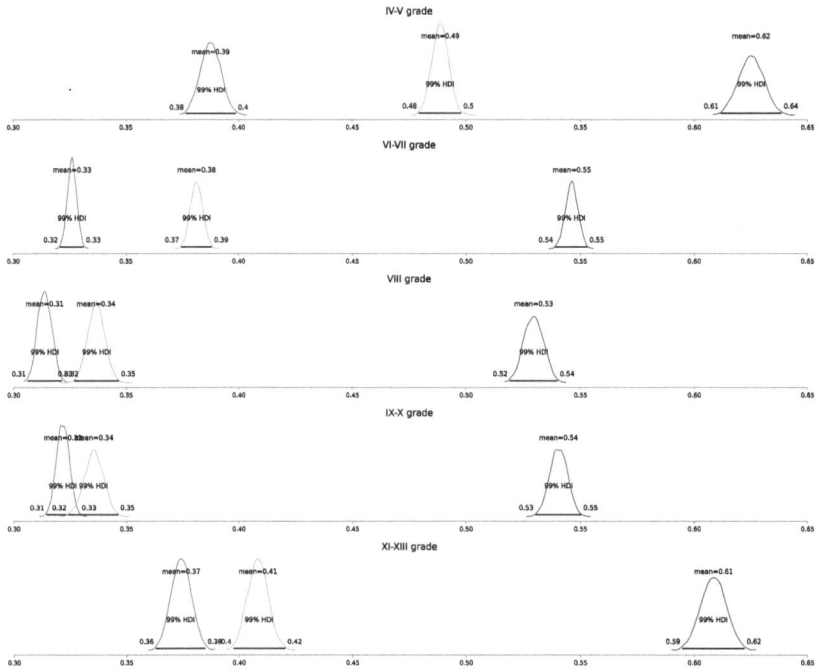

Fig. 2. Diagrams showing the differences between actual data ($p_{singles}$ blue, $p_{doubles}$ orange) and fully successful collaboration (p_{best} green); worst-case collaborations are not shown. (Color figure online)

Figure 2 shows the distribution of probabilities for the five age categories for the typical teams (see 2.2). We can observe that the improvement from $p_{singles}$ to $p_{doubles}$ is greater for the youngest and decreases with age. Such an improvement is less evident in the categories where $p_{singles}$ is already low; for IX-X grade, there is even some uncertainty about the actual improvement. The gap between $p_{doubles}$ and the best-case benchmark is large for most categories, except for IV–V grade.

3.2   Impact of Tasks Content and Features

For each task t of the 60 tasks used in the contest, we computed the probability distribution of the event "correctly solving task t" for singles and doubles, and the delta between them. The diagram in Fig. 3 positions all 60 tasks considering the mean probability for singles ($p_{singles}$) and the delta $p_{doubles} - p_{singles}$. The more a task is on the right (high $p_{singles}$), the simpler it resulted (high success probability by a single solver); the more a task is in the upper part of the diagram (high delta), the higher was the increase of success probability from single to double teams.

The red points are those where the delta is positive with high certainty (99% HDI of $p_{singles}$ and $p_{doubles}$ do not overlap), whereas the gray ones are those where the evidence of an increase is not certain enough (99% HDI of $p_{singles}$ and $p_{doubles}$ partially overlap, even if sometimes just slightly).

It is worth noticing that the top left part of the diagram is empty; this means that for hard tasks there is no benefit from working in pairs. Similarly, the most improvement is measured more often on tasks of medium difficulty. We analyzed some of the tasks where the improvement was more evident and we made some hypotheses about the content and features of those task, that would explain such improvements. Due to space constraints, here we discuss only one of them.

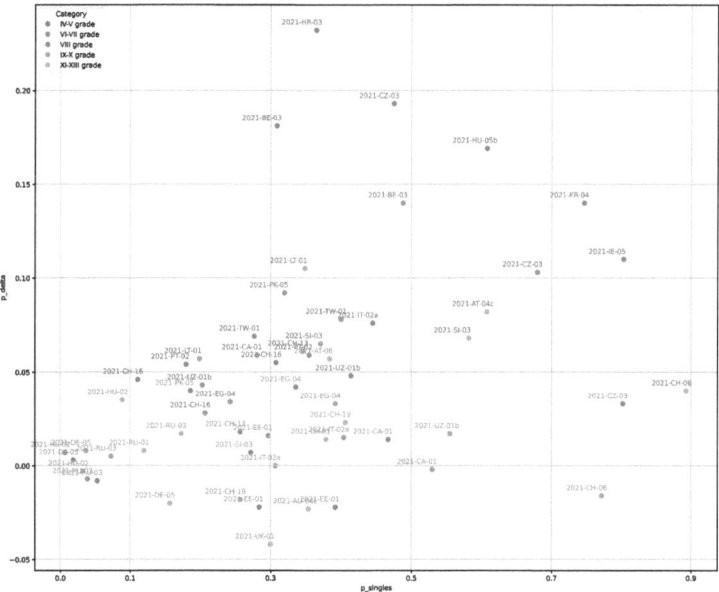

Fig. 3. Scatterplot of the tasks according to the success probability for the task and the size of improvement on the task moving from singles to doubles. For red tasks the increase has a solid evidence in the estimation (99% HDIs of $p_{singles}$ and $p_{doubles}$ do not overlap); for gray tasks 99% HDIs of $p_{singles}$ and $p_{doubles}$ overlap. The color of the dot gives the age group. (Color figure online)

Task 2021-BE-03 (Necklaces instruction). This task proposes a programming exercise requiring to write a sequence of characters complying with a given syntax. We make the hypothesis that doubles are better able to note and correct syntax errors that could go unnoticed to an individual. This hypothesis is in line with the literature, where the fact that syntax is a hurdle in learning to program is often discussed and where this fact has motivated the development of block-based programming languages, the use of Parsons problems, *etc.*. We analyzed the data for the "IV-V grade" age group, where the collaboration was more effective. We collected the following measures relevant for the solving process:

Total number of modifications. The times the solvers modified the string of characters: it happened on average 26.7 (std. dev.: 23.5) times for singles, 27.1 (std. dev.: 20.7) times for doubles.

Number of corrections. The times the solvers modified the string of characters excluding appends (these insertions or changes are likely to be corrections): it happened on average 5.6 (std. dev.: 9.5) times for singles, 5.7 (std. dev.: 10.5) times for doubles.

Percent number of corrections. The times the solvers modified the string of characters excluding appends w.r.t. total modifications: it is on average 13.7% (std. dev.: 0.14) for singles, 14.1% (std. dev.: 0.13) for doubles.

Number of resets. The times the solvers deleted the string of characters: it happened on average 0.8 (std. dev.: 1.5) times for singles, 0.7 (std. dev.: 1.3) times for doubles.

Percent times the solution was changed into wrong. How frequently was a correct solution then changed into a wrong one: it happened for 0.8% of the singles and for 0.4% of the doubles. Overall it happened only for 0.5% of the teams.

The two populations differ somewhat and the doubles seem to be slightly more active with the platform, but no macroscopic differences were found. This task however shows a remarkable property: solvers are in general very stable on a correct solution, when it is found. In other words, a correct solution is easy to recognize as such. This could explain why the doubles improved so much (the success ratio is 31% for singles and 49% for doubles): it is enough that one of the two solvers identifies the correct solution, the other will accept it easily; in fact $p_{double} = 0.49$ is very close to $p_{best} = 0.52$. In order to check the validity of this last observation we analyzed also the data for 2021-EE-01, a task in which, in the same "IV–V grade" age group the increment for doubles is dubious. The "Percent times the solution was changed into wrong" for 2021-EE-01 is much higher than for 2021-BE-03: 22% for singles and 26% for doubles, 25% overall.

4   Limitations and Threats to Validity

Indirect measures of collaboration. The main limitation of this study is that we do not have any direct data about how teams solve tasks and collaborate. We only have the measure of their performance in the Bebras challenge, and some indirect data provided by log data related to their interaction with the Bebras platform during the contest. Thus, our findings cannot be considered definitive, and need to be further checked, *e.g.,* possibly with in-depth qualitative study based on observing students interacting with a mate when solving tasks. However, the size of the data set and the rigorous methods used to analyze it supports the validity of these preliminary findings, which suggest promising directions for future investigations.

Independence of team size from team ability. If doubles were formed by pupils with lower prior ability, this would provide some explanation for the limited improvement observed between the performance of doubles w.r.t. singles. We addressed this possible bias in two ways. On the one hand we excluded from the analysis the 8'706 teams whose composition type resulted atypical w.r.t. the rest of teams of their teachers. The average increment from singles to doubles computed on the original dataset results to be slightly lower than the one showed in Table 1. The inclusion of the small proportion of teams (8%) that are possibly biased w.r.t. ability decreases slightly the advantage of having a second person in the team, supporting the hypothesis that their teachers assigned the best students to the single teams. On the other hand, we studied the geographic provenance of teams and used this as a proxy for their ability; more precisely, we used the results of standardized tests conducted every year in all schools of our country[5]. We found neither evident trends nor correlations between the proportion of singles in a region and the results in standardized tests in that region, which suggests that there is no correlation between the prior ability of teams and their composition. Even though the test results are available also with finer definition (*e.g.,* by individual school), we conducted our analysis only at the regional level, since it is not mandatory for registered teams to enter details about their school. Moreover, an analysis at the school level would pose several legal problems since we did not ask in advance for an explicit consent and the data about the standardized tests are not available as open data.

Source for team size data. The team size for each team is entered by teachers when they register their teams, and we have no direct control on the fact that the actual size of a team corresponds to the declared one. In particular many situations may occur (*e.g.,* absence of a mate the day of the contest, odd number of pupils in a class, …) that yield to a team registered as doubles actually being formed by a single student only. However, in order to produce certificates for their teams after the challenge, teachers had the possibility to enter in the system additional information on the teams' members. Most teachers used this feature. In order to address the possible bias of false doubles, we did not used the declared team type but adjusted the team type value in our dataset as follows: i) we excluded from the analysis all teams without explicit information on their members, since it is dubious whether they should be indeed considered as doubles or singles; ii) similarly, we set the team size type according to the number of filled in members (in some cases this meant to change the composition w.r.t. the one

---

[5] We used the data provided by INVALSI for the school year 2021, taken from https://invalsi-serviziostatistico.cineca.it/; see also [11] for a previous study.

declared upon teams registration). As a result we ended up considering a dataset of 19'490 teams, among the larger number of 28'196 teams who participated in the challenge. The remaining 8'706 teams are not invalid, but their size was uncertain and we preferred to restrict the analysis to data with some guarantees to have been curated by the teachers.

Contest aggregation. One could also take into account that the tasks come packed together in a suite of twelve. We carried out the same analysis by starting with a model with contest data aggregated by suites, but we did not find any visible difference. In principle the data observed on suites could fit the model worse (note that the two models are mathematically equivalent). The difference in the uncertainty is negligible, therefore considering the tasks independent one from each other seems to be a viable hypothesis.

5   Related Work

Group work is often proposed as a way for improving learning, and many studied the social and emotional advantages children can gain from working together [2]. In particular, collaborative learning is an educational approach to teaching and learning that involves groups of learners working together to solve a problem, complete a task, or create a product [4]. However, while collaboration in pairs or small groups can facilitate pupils' learning and development, many observations of classroom practice show that group work does not realise the potential promised by research [10]. Sometimes peer interaction can even result in poorer learning outcomes [6]. In fact, although collaboration is often considered a beneficial learning strategy, identifying the key factors which make a collaboration successful or not is still an open issue. [7] studied important features for educators to consider when deciding when and how to include collaboration in instructional activities. Our study tries to understand in which context or task the collaboration is more effective. In 2015 the Programme for International Student Assessment (PISA[6]) launched the first large-scale, international assessment to evaluate students' competency in collaborative problem solving. It required students to interact in order to solve problems. It included group decision-making tasks (requiring argumentation, debate, negotiation or consensus to arrive at a decision), group co-ordination tasks (including collaborative work), and group-production tasks (where a product must be created by a team, including designs for new products or written reports). Collaborative problem-solving performance is positively related to performance in the core PISA subjects (science, reading, and mathematics), but the relationship is weaker than that observed among those domains. Girls perform significantly better than boys in collaborative problem solving in every country and economy that participated in the assessment; students have a generally positive attitude towards collaboration [8].

---

[6] See https://www.oecd.org/pisa.

6   Conclusions

Our observational study confirms that the effect of collaboration is positive, but it is rather limited compared to what one could expect from a fully successful collaboration. The positive effect of collaboration seems somewhat to decrease when the grade increases: this certainly needs further in-depth analysis, it could be related to some specificity of the age groups, but also to task features, often rather different for older students. For example, when a correct solution is easy to recognize, the collaboration seems to work more efficiently. The main limitation of this study is that we did not directly observe how teams solved tasks and collaborated. Even the interaction data we analyzed are indirect and can be interpreted in different ways. Our findings, although promising, should be considered preliminary and we intend to design a follow up qualitative study, in which we will observe students interacting to solve tasks.

References

1. Dagienė, V.: Sustaining informatics education by contests. In: Hromkovič, J., Královič, R., Vahrenhold, J. (eds.) ISSEP 2010. LNCS, vol. 5941, pp. 1–12. Springer, Heidelberg (2010). https://doi.org/10.1007/978-3-642-11376-5_1
2. Galton, M., Williamson, J.: Groupwork in the Primary Classroom. Routledge, London (2003)
3. Haberman, B., Cohen, A., Dagienė, V.: The beaver contest: attracting youngsters to study computing. In: Proceedings of ITiCSE 2011, pp. 378–378. ACM, Darmstadt (2011)
4. Laal, M., Ghodsi, S.M.: Benefits of collaborative learning. Procedia-Soc. Behav. Sci. **31**, 486–490 (2012)
5. McElreath, R.: Statistical Rethinking: A Bayesian Course with Examples in R and Stan. Chapman and Hall/CRC, New York (2020)
6. Messer, D.J., Joiner, R., Loveridge, N., Light, P., Littleton, K.: Influences on the effectiveness of peer interaction: children's level of cognitive development and the relative ability of partners. Soc. Dev. **2**(3), 279–294 (1993)
7. Nokes-Malach, T.J., Richey, J.E., Gadgil, S.: When is it better to learn together? Insights from research on collaborative learning. Educ. Psychol. Rev. **27**(4), 645–656 (2015). https://doi.org/10.1007/s10648-015-9312-8
8. OECD: PISA 2015 Results (Volume V) (2017)
9. Williams, L.: Integrating pair programming into a software development process. In: Proceedings of the 14th Conference on Software Engineering Education and Training, pp. 27–36 (2001). https://doi.org/10.1109/CSEE.2001.913816
10. Wood, D., O'Malley, C.: Collaborative learning between peers. Educ. Psychol. Pract. **11**(4), 4–9 (1996). https://doi.org/10.1080/0266736960110402
11. Bellettini, C., Lonati, V., Monga, M., Morpurgo, A.: An analysis of the performance of Italian schools in Bebras and in the national student assessment INVALSI. In: Fronza, I., Pahl, C. (ed.) Proceedings of the 2nd Systems of Assessments for Computational Thinking Learning Workshop (TACKLE 2019). CEUR Workshop Proceedings, vol. 2434 (2019)
12. Bellettini, C., Lonati, V., Monga, M., Morpurgo, A.: Behind the shoulders of bebras teams: analyzing how they interact with the platform to solve tasks. In: Lane, H.C., Zvacek, S., Uhomoibhi, J. (eds.) CSEDU 2019. CCIS, vol. 1220, pp. 191–210. Springer, Cham (2020). https://doi.org/10.1007/978-3-030-58459-7_10

Assessing Computational Thinking: The Relation of Different Assessment Instruments and Learning Tools

Vaida Masiulionytė-Dagienė$^{(\boxtimes)}$ ⓘ and Tatjana Jevsikova$^{(\boxtimes)}$ ⓘ

Vilnius University Institute of Data Science and Digital Technologies, Akademijos street 4, 08412 Vilnius, Lithuania

{vaida.masiulionyte-dagiene,tatjana.jevsikova}@mif.vu.lt

Abstract. The relevance of computational thinking as a skill for today's learners is no longer in question, but every skill needs an assessment system. In this study, we analyze two validated instruments for assessing computational thinking - the CTt (Computational Thinking Test) and the CTS (Computational Thinking Scale). The study involved 49 students in grades 8 and 9 (age 14–16). Prior to the study, students in both grades were taught computational thinking differently. One group learned computational thinking by completing tasks and creating projects in Scratch, the other group learned by completing tasks in "Minecraft: Education Edition". The students were asked to take the CTt and CTS tests. The nature of these tests is different, one is computational thinking diagnostic tool, the other is a psychometric self-assessment test consisting of core abilities (subconstructs) important for computational thinking. The aim of this study was to determine how these tests related to each other and whether students' gender and the different tools chosen to teach computational thinking had an impact on the level of computational thinking knowledge and abilities acquired based on the tests. The results have shown that the scores of the two tests correlated with each other only for male students' subgroup. For a whole group CTt scores correlated only with CTS algorithmic thinking subconstruct. The results have also shown that teaching tools do have an impact on the acquisition of different computational thinking concepts skills: students taught with different tools had different test results. This study provides useful implications on computational thinking teaching improvement and its assessment better understanding.

Keywords: Computational thinking · Assessment · Computational thinking assessment instruments · CTt · CTS · Gender differences · Learning tools

1 Introduction and Background

Computational thinking education is an important component in the process of digitalization of society and the economy, as discussed in a recent study organized by the European Commission [2]. The European Commission encourages this focus by making computational thinking education a priority in order to improve digital skills and

A. Bollin and G. Futschek (Eds.): ISSEP 2022, LNCS 13488, pp. 66–77, 2022.
https://doi.org/10.1007/978-3-031-15851-3_6

competences as part of the digital transformation. This study highlights that the transfer of computational thinking into educational subjects is a new field, which poses many challenges that need to be assessed in educational practice.

Evaluation and measurement of results are important in educational practice when implementing different teaching methods and tools. In the absence of a single unified definition of computational thinking, researchers working in this area apply different definitions [19, 25, 26]. As the result, there is also no single tool to assess computational thinking. For the assessment of computational thinking, various methods and tools have been developed today, such as Dr. Scratch, Bebras tasks, Zoombinis, CTt, CTS, etc.

Román-González categorized the tools for assessing computational thinking into the following groups: diagnostic tools, summative tools, formative-iterative tools, data-mining tools, skill transfer tools, perceptions-attitudes scales, vocabulary assessment [20].

Dr. Scratch as formative-iterative tool automatically analyzes Scratch programming projects and also can be used to develop computational thinking [27, 28]. Dr. Scratch analyzes code based on these computational thinking concepts: abstraction and problem decomposition, logical thinking, synchronization, parallelism, algorithmic notions of flow control, user interactivity, data representation [16].

Bebras tasks are used as another computational thinking assessment tool, classified as skill transfer tool [4, 5, 14, 20]. It is mentioned that the Bebras challenge tasks refer to analytics or analytical thinking concept [14], but there is not yet a representative set of Bebras tasks that has been validated as an assessment instrument for computational thinking.

Zoombinis [1] is an award-winning educational game from the nineties that has been rebuilt for modern platforms. It is not only used for learning computational thinking, but also in recent years for computational thinking assessment as a data-mining assessment tool. In Zoombinis, all the players actions are logged and then analyzed for the purpose of learning or the assessment. Concepts that are assessed in Zoombinis: problem decomposition, pattern recognition, abstraction, algorithm design [21].

More approaches to computational thinking assessment appear in recent research, e.g., CT-cube, a framework for the design, realization, analysis, and assessment of computational thinking activities [17], data driven approaches based on students' artefacts [6].

CTt (computational thinking test) was developed and validated by Román-González et al. [19]. CTt consists of 28 questions, divided in 7 groups: basic directions and sequences; repeat times; repeat until; simple conditional; complex conditional; while conditional; simple functions [19]. The test focuses on middle school children (mainly for 12–14 years old, but it can be used from 5th to 10th grade). Computational concepts used in the test are aligned with the CSTA (Computer Science Teachers Association) Computer Science Standards for the 7th and 8th grade [3]. Guggemos [9] mentions the main computational thinking concepts that CTt covers: abstraction, decomposition, algorithms, and debugging. The CTt has some advantages, like the ability to be conducted in large groups in pre-test scenarios, allowing for early detection of students with high abilities (or special needs) for programming tasks; and the ability to collect quantitative information before the evaluations of the effectiveness of curricula designed to foster computational thinking [19].

Another validated tool for assessing computational thinking independently of programming is the CTS (Computational Thinking Scales) test [13]. This test identifies the following components of computational thinking based on ISTE (International Society for Technology in Education): creativity, algorithmic thinking, critical thinking, problem solving and collaboration skills [12]. The specificity of this test is that it is a self-assessment instrument. In this respect, it is completely different from the CTt test, which is knowledge assessment test.

Since computational thinking is broader than programming as defined in The European Commission's Staff Working Document accompanying the Digital Education Action Plan 2021–2027 (DEAP) [7]: "Computational thinking, programming and coding are often used interchangeably in education settings, but they are distinct activities", therefore it is important to pay attention to computational thinking assessment tools that assess beyond programming or algorithmic skills. Commonly used validated tests for assessing computational thinking that are independent of programming languages are the CTt and CTS [10, 15, 18, 24]. Other tools mentioned above are associated with specific platforms for programming or gaming (e.g. Dr. Scratch, Zoombinis). The Bebras tasks are tool independent, but not yet validated as a set of tasks for the assessment of computational thinking. For this reason, the tool-independent tests mentioned above (CTt and CTS) were chosen for this study. Both CTt and CTS are presented to students in a form of tests (a set of questions/statements with a set of answer options), while allowing to assess computational thinking from different points of view.

Due to the complexity of computational thinking, it is also relevant to assess computational thinking in a complex way, using more than one tool or method [22, 28]. According to the beforementioned classification [20], the CTt test falls into the diagnostic tools group and the CTS test into the perceptions-attitudes scales group [20]. Guggemos et al. [10] also mention that a system of different assessments is essential because various computational thinking profiles can be identified using multifunctional methods. Guggemos et al. [10], in their research for the assessment of computational thinking, analyze these two tests: CTt and CTS as well. However, researchers do not consider the computational thinking teaching tools used, nor do they differentiate between students' computational thinking test scores according to their gender.

The *aim* of this study is to investigate the relationship between CTt and CTS tests in relation to students' gender and the tools used to develop computational thinking.

We pose the following *research questions*:

RQ1: Is there a relationship between students' CTt and CTS (including its subconstructs) scores?
RQ2: How students' CTt and CTS tests' results are associated with the learning tool used to develop CT and differ in gender groups?

The paper is structured as follows. First, we present learning methods and tools, describe respondents, instruments and data analysis methods. Next, we present the results of the study according to the research questions. Finally, we discuss our findings, describe limitations and provide directions for future research.

2 Methods

2.1 Learning Tools and Methodology

Students were taught computational thinking using Scratch and "Minecraft: Education Edition". The teaching tools were not a part of the study, the computational thinking knowledge was acquired during the regular computer science lessons using the above-mentioned tools. Each class was familiar with both tools, but the main tool for grade 8 was the Scratch platform, while grade 9 learned in the "Minecraft: Education Edition" environment. On the Scratch platform, the students had to do "open" tasks using computer science concepts, such as cycles, conditions: they wrote programs that draw different shapes, developed a project with self-created characters, environment and implemented a created scenario of interaction between the characters. On the "Minecraft: Education Edition" platform, students learnt from the pre-designed lessons based on CSTA and ISTE guidelines [3, 12]. They first completed the tasks from the block programming fundamentals lessons, based on the 5 lessons provided, and then were introduced to the basics of Python programming (also completing the tasks from the 5 lessons). One lesson in "Minecraft: Education Edition" required 1 or 2 academic lessons to complete all the activities. On average, both grades had 12–14 lessons using these tools. As all tasks were completed individually and there were no team tasks during this learning period, the concept of cooperativity was removed from the CTS test in this study.

2.2 Respondents

In total, 49 students (51% female and 49% male), studying in school grades 8 and 9 (aged 14–16), took part in the survey. There were 24 students of 8th grade (51%), learning computational thinking with Scratch as dominating tool, and 25 students of 9th grade (49%), learning with "Minecraft: Education edition" as a primary tool.

All respondents were informed of the purpose of the study and gave their free will consent to participate in the study.

2.3 Instruments

In this study, besides the questions on basic demographical information, the two following instruments were used.

CTt. A validated instrument, consisting of 28 questions [19]. CTt test is claimed to be unidimensional [10] although addressing 7 cognitive operations (4 items for each cognitive operation arranged in increasing difficulty direction): basic directions and sequences, loops repeat times, loops repeat until, if simple conditional statement, if/else complex conditional statement, while conditional, and simple functions. For each question, four answer options are suggested with only one correct. Each item is rated as 1 (correct) or 0 (incorrect).

CTS. A validated computational thinking assessment scale, originally consisting of creativity, algorithmic thinking, critical thinking, problem solving and cooperativity subconstructs, rated on a 5-point Likert scale [13]. In our study, we included all subconstructs of this scale except for cooperativity, as mentioned before.

2.4 Data Analysis

For the analysis of the collected data, quantitative methods were used. Data normality for a whole sample has been checked with Kolmogorov-Smirnov and Shapiro-Wilk tests. CTt scores were not normally distributed. Due to this reason as well as analysis involving relatively small subgroup analysis, we used distribution-free non-parametric measures:

- To compare differences between two independent samples, the Mann–Whitney U test was used, and η^2 was used as an effect size measure.
- To test the monotonous relationships between the pairs of variables, Spearman's rank correlations were used.

We computed scores of the tests and their parts as a sum of the item scores.

The reliability of CTS psychometric scale subconstructs was examined using Cronbach's Alpha. After evaluating subscale reliability, item 4 from problem solving subconstruct was dropped to improve subscale reliability. Cronbach's Alphas for scale subconstructs were satisfying (≥ 0.7): 0.701 for creativity (8 items), 0.765 for algorithmic thinking (6 items), 0.725 for critical thinking (5 items), 0,703 for problem solving (5 items).

The significance level was set to $\alpha = 0.05$.

For the statistical analysis, IBM SPSS Statistics 28 software package and MS Excel were used.

3 Results

3.1 An Association of Students' CTt and CTS Results

In a whole group of students, the CTt scores ranged from 9 to 27 with mean scores of 21.2, while CTS scores ranged from 57 to 107 with mean value of 77.4. The descriptive statistics for the results of both tests in general and according to tests' subscales, are presented in Table 1.

Table 1. Descriptive statistics for test scores (N = 49).

Cognitive operation/construct	Score range	Min.	Max.	Mean	Std. deviation
CTt	0–28	9	27	21.2	4.5
Sequences	0–4	2	4	3.7	0.5
Loops (times)	0–4	0	4	3.5	0.8
Loops (until)	0–4	1	4	3.2	0.9
If (simple)	0–4	0	4	2.7	1.1
If (complex)	0–4	0	4	2.6	1.1
While	0–4	0	4	2.5	1.3
Functions	0–4	0	4	3.0	1.2
CTS	30–150	57	107	77.4	10.0
Creativity	8–40	19	39	28.2	4.0
Algorithmic thinking	6–30	10	24	17.4	3.5
Critical thinking	5–25	5	22	15.0	3.2
Problem solving	5–25	12	23	16.8	2.6

Table 2. Spearman's rank correlations between CTS and its subconstructs and CTt scores.

	CTS	Creativity	Algorithmic thinking	Critical thinking	Problem solving
ρ	0.174	0.000	0.307*	−0.034	0.212
p	0.232	1.000	0.032	0.817	0.143

* Significant at 0.05 level (2-tailed)

Spearman's rank correlations for 49 students have been calculated between CTt scores and CTS general scores as well as its subconstructs (Table 2).

Significant (at 0.05 level) relationship was found between CTt and algorithmic thinking scores ($\rho = 0.307$, $p = 0.032$). However, there was no significant association between CTt scores and CTS general results ($\rho = 0.174$, $p = 0.232$). Further analysis on differences between students' groups is presented in the following section.

3.2   CT Assessment Scores in Learning Tool and Gender Groups

In order to observe the relationships between CTt scores and CTS including its subconstruct scores between groups studying with different learning tool (Minecraft: Education Edition, Scratch) and between male and female students' subgroups, Spearman's rank correlations have been computed (Table 3). We use Minecraft as a shortened name form of "Minecraft: Education Edition".

Table 3. Spearman's rank correlations between CTS and its subconstructs and CTt scores for Minecraft and Scratch learning groups and gender.

Group	Measure	CTS	Creativity	Algorithmic thinking	Critical thinking	Problem solving
Minecraft	ρ	0.371	0.216	0.414*	0.206	0.259
(n = 25)	p	0.068	0.299	0.040	0.323	0.211
Scratch	ρ	0.285	0.103	0.301	0.128	0.307
(n = 24)	p	0.178	0.631	0.154	0.550	0.144
Males	ρ	0.445*	0.301	0.576**	0.119	0.276
(n = 24)	p	0.029	0.153	0.003	0.581	0.191
Females	ρ	−0.033	−0.188	0.137	−0.138	0.159
(n = 25)	p	0.877	0.367	0.514	0.510	0.447

* Significant at 0.05 level (2-tailed); ** significant at 0.01 level (2-tailed)

We found a significant monotonous relationship between CTt and CTS scores in a subgroup of male students ($\rho = 0.445$, p = 0.029). However, there were no correlations between scores in other subgroups (female students or subgroups based on learning tool). The strongest significant monotonous relationship was found between CTS's algorithmic thinking and CTt scores in a group of boys ($\rho = 0.576$, p = 0.003). In a group of students learning with Minecraft as a primary tool, this relationship was also significant, but weaker ($\rho = 0.414$, p = 0.040).

Graphically, the differences in CTt scores between groups studied are presented in Fig. 1.

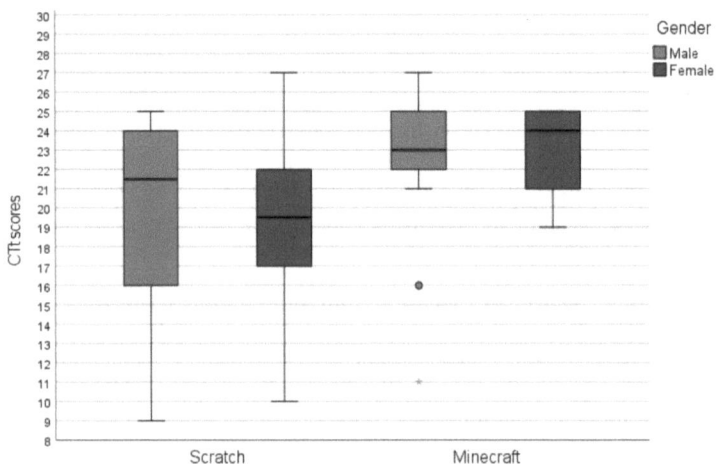

Fig. 1. CTt scores for Scratch and Minecraft groups' male and female students.

The mean ranks of scores for different test constructs and subconstructs in groups studied and results of the Mann-Whitney U tests are presented in Table 4 (light grey shading for significance at 0.05 level, dark grey for significance at 0.01 level).

Table 4. Differences between groups (Scratch, Minecraft, male and female): Mann-Whitney U tests' results.

Cognitive opera-tion/Construct	Mean rank		Z	p	Mean rank		Z	p
	Scratch	Minecraft			Males	Females		
CTt total scores	20.44	29.38	-2.20	0.028	25.67	24.36	-0.32	0.748
Sequences	24.40	25.58	-0.40	0.690	26.35	23.70	-0.89	0.371
Loops Times	24.23	25.74	-0.43	0.665	24.38	25.60	-0.35	0.725
Loops Until	21.38	28.48	-1.87	0.061	25.15	24.86	-0.08	0.940
If simple	19.67	30.12	-2.68	0.007	25.25	24.76	-0.13	0.900
If complex	23.21	26.72	-0.91	0.363	21.17	28.68	-1.95	0.052
While	20.52	29.30	-2.22	0.027	26.35	23.70	-0.67	0.502
Functions	24.27	25.70	-0.38	0.705	26.65	23.42	-0.86	0.392
CTS total scores	28.27	21.86	-1.57	0.116	26.56	23.50	-0.75	0.453
Creativity	28.35	21.78	-1.62	0.106	23.85	26.10	-0.55	0.581
Algorithmic thinking	25.71	24.32	-0.34	0.732	29.83	20.36	-2.34	0.020
Critical thinking	30.92	19.32	-2.86	0.004	25.10	24.90	-0.05	0.960
Problem solving	26.23	23.82	-0.60	0.551	24.71	25.28	-0.14	0.887

The Mann-Whitney U test confirmed a significant difference in CTt scores between the groups learning as a primary tool with Scratch (mean rank 20.44) and Minecraft (mean rank 29.38): Z = –2.20, p = 0, 028. An effect size $\eta^2 = 0.1$ denotes that 10% of variance in rank was accounted by the CT learning tool used (Scratch or Minecraft).

Significantly higher scores in critical thinking were observed in Scratch group compared to Minecraft (Z = –2,86, p = 0.004, $\eta^2 = 0.17$). Interquartile range of the differences in critical thinking scores, including subgroups of male and female students are presented graphically in Fig. 2.

Significant differences for Scratch and Minecraft groups were also found in scores of CTt cognitive operation "simple If" (Z = –2.68, p = 0.007, $\eta^2 = 0.15$) and "While loop" (Z = –2.22, p = 0.027, $\eta^2 = 0.1$).

Studying the differences between groups of boys and girls, significant differences were found in algorithmic thinking scores: Z = –2.34, p = 0.020, $\eta^2 = 0.12$, with mean rank for female students 20.36 and for male students 29.83.

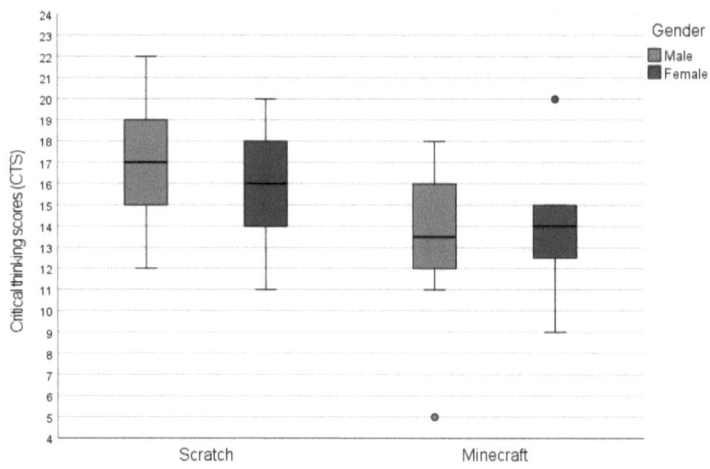

Fig. 2. Critical thinking (CTS) scores for Scratch and Minecraft groups' male and female students.

4 Discussion and Conclusion

In this study, we examined an association between the results gained by the two computational thinking assessment instruments and differences in scores based on groups of learning tool used to develop computational thinking and gender.

4.1 Relationship Between Students' CTt and CTS Scores

Looking at the results regarding the first research question, there was no significant relationship found between CTt and CTS general scores. This finding is in line with the recent study by Guggemos et al. [10] and can be explained by different nature of the instruments. However, it is interesting to note that analysis of the results of both tests in separate groups of students by gender has shown that the tests' scores correlated with each other in the group of male students.

Analysis, performed on the results for the individual subconstructs of the CTS test, we also see monotonous positive relationship of the CTt scores with the algorithmic thinking scores in CTS test, what supports the results of the study by Guggemos et al. [10]. Thus, in response to the first research question, we can say that the tests do not correlate from a generic point of view, but that the correlation is influenced by the gender of the students, and to fully validate this statement, it would require research in a larger group of students.

4.2 Differences in the CTt and CTS Tests' Results in Students' Groups by Learning Tool and Gender

In response to the second research question, which asks how the test results were influenced by the teaching tools, we can see that students who studied using "Minecraft: Education Edition" had significantly better CTt test results than those students who studied computational thinking in Scratch. These results could be explained by the fact that both "Minecraft: Education Edition" and CTt are based on the CSTA [3] teaching standards. The pre-designed lessons include the same elements as the CTt test (cycles, conditional sentences, etc.). This is further confirmed by the results of the separate CTt test groups, cognitive operations such as the "while loop" and the "simple if" conditions. However, students who had studied in the Scratch environment had better results on the critical thinking subconstruct of the CTt test. Critical thinking is defined as "the use of cognitive skills or strategies that increase the possibility of the desired behaviors" [11]. In the context of the definition of critical thinking, the results obtained can be explained by the fact that, unlike the pre-prepared lessons used in "Minecraft: Education Edition", in the Scratch environment, students had to create their own projects and find custom solutions to achieve the desired outcome. As we can see, different tools develop different computational thinking skills during the teaching process, and this should be considered when teaching computational thinking, so that the most versatile computational thinking skills can be developed and assessed with the widest possible range of assessment tools. There is also a need for more research in this area on which tools best develop which computational thinking skills.

In terms of gender, the significant difference was observed in the algorithmic thinking subconstruct of the CTS test. Boys showed higher scores in algorithmic thinking than girls. On a one hand side, this finding reflects the existing stereotype of computer science and engineering being more male-oriented field [15, 23]. While the findings of Ma et al. [15] study show that the CTS test algorithmic thinking scores before the intervention were slightly higher in the girls' group, after the intervention, the scores became identical, with a non-significant difference to the boys' advantage. As Groher et al. [8] mentions: "diversity among the students calls for diversity among the teaching and learning materials." This could be one of the reasons for the different tests results.

4.3 Limitations and Future Research Directions

The main limitation of this study was the relatively small sample size of students involved and obtained by the convenient sampling method. Also, the slightly different age group of the students (8th and 9th grade) might have had some influence on the results. Nevertheless, the results were in line with the findings of other related studies and provided interesting insights for further research with greater samples and other methods.

In addition to the test results, one trend was observed during the course of the study that would allow for improvements in the assessment. When taking the test, students used their hand or a computer mouse to guide the screen through the picture next to each test question in order to find the correct answer. However, this process was not logged anywhere and we only saw one of the selected answers as the test result. However, in order to assess computational thinking, we should assess the process of thinking itself. Such approach we may see in the Zoombinis game [21]. It might be possible that the student's thinking process was partly correct, e.g. right at the beginning with a slight mistake at the end, resulting in the wrong answer, but this is not what we see as a test score for diagnostic test. Such approach could also help to eliminate the cases when the student clicked the right answer by chance. In future research, we will focus on how to better assess the process thinking for the task solution, and not just the final result.

References

1. Asbell-Clarke, J., et al.: The development of students' computational thinking practices in elementary- and middle-school classes using the learning game. Zoombinis. Comput. Hum. Behav. **115**, 106587 (2021)
2. Bocconi, S., et al.: Reviewing computational thinking in compulsory education. In: Inamorato Dos Santos, A., Cachia, R., Giannoutsou, N., Punie, Y. (eds.) Publications Office of the European Union, Luxembourg (2022). https://doi.org/10.2760/126955. ISBN 978-92-76-47208-7, JRC128347
3. CSTA: K12 computer science standards (2017). https://www.csteachers.org/page/about-csta-s-k-12-nbsp-standards
4. Djambong, T., Freiman, V., Gauvin, S., Paquet, M., Chiasson, M.: Measurement of computational thinking in K-12 education: the need for innovative practices. In: Sampson, D., Ifenthaler, D., Spector, J., Isaías, P. (eds.) Digital Technologies: Sustainable Innovations for Improving Teaching and Learning, pp. 193–222. Springer, Cham (2018). https://doi.org/10.1007/978-3-319-73417-0_12
5. Dolgopolovas, V., Jevsikova, T., Dagiene, V., Savulioniene, L.: Exploration of computational thinking of software engineering novice students based on solving computer science tasks. Int. J. Eng. Educ. **32**(3), 1–10 (2016)
6. Eloy, A., Achutti, C.F., Cassia, F., Deus Lopes, R.: A data-driven approach to assess computational thinking concepts based on learners' artifacts. Inf. Educ. **21**(1), 33–54 (2022)
7. European Commission: Digital Education Action Plan 2021–2027: Resetting education and training for the digital age (2020). https://eurlex.europa.eu/legal-content/EN/TXT/?uri=CELEX:52020DC0624
8. Groher, I, Sabitzer, B., Demarle-Meusel, H., Kuka, L., Hofer, A.: Work-in-progress: closing the gaps: diversity in programming education. In: 2021 IEEE Global Engineering Education Conference (EDUCON), pp. 1449–1453 (2021)
9. Guggemos, J.: On the predictors of computational thinking and its growth at the high-school level. Comput. Educ. **161**, 104060 (2021)
10. Guggemos, J., Seufert, S. Román-González, M.:Computational thinking assessment – towards more vivid interpretations. Tech. Know. Learn.https://doi.org/10.1007/s10758-021-09587-2
11. Halpern, D.F.: Thoughts and Knowledge: An Introduction to Critical Thinking. Lawrence Erlbaum Associates, New Jersey-London (1996)
12. ISTE: Computational thinking: leadership toolkit (2015). https://www.iste.org/computational-thinking
13. Korkmaz, Ö., Çakir, R., Özden, M.Y.: A validity and reliability study of the computational thinking scales (CTS). Comput. Hum. Behav. **72**, 558–569 (2017)
14. Labusch, A., Eickelmann, B.: Computational thinking competences in countries from three different continents in the mirror of students' characteristics and school learning. In: Kong, S.C., et al., (eds.) Proceedings of International Conference on Computational Thinking Education 2020, pp. 2–7. The Education University of Hong Kong (2020)
15. Ma, H., Zhao, M., Wang, H., Wan, X., Cavanaugh, T.W., Liu, J.: Promoting pupils' computational thinking skills and self-efficacy: a problem-solving instructional approach. Educ. Tech. Res. Dev. **69**(3), 1599–1616 (2021). https://doi.org/10.1007/s11423-021-10016-5
16. Moreno-León, J., Robles, G., Román-González, M.: Dr. Scratch: automatic analysis of scratch projects to assess and foster computational thinking. RED-Rev. Educ. Distancia **46**, 1–23 (2015)
17. Piatti, A., et al.: The CT-cube: a framework for the design and the assessment of computational thinking activities. Comput. Hum. Behav. Rep. **5**, 100166 (2022)

18. Poulakis, E., Politis, P.: Computational thinking assessment: literature review. In: Tsiatsos, T., Demetriadis, S., Mikropoulos, A., Dagdilelis, V. (eds.) Research on e-Learning and ICT in Education, pp. 111–128. Springer, Cham (2021). https://doi.org/10.1007/978-3-030-643 63-8_7

19. Román-González, M., Pérez-González, J.-C., Jiménez-Fernández, C.: Which cognitive abilities underlie computational thinking? Criterion validity of the computational thinking test. Comput. Hum. Behav. **72**, 678–691 (2017)

20. Román-González, M., Moreno-León, J., Robles, G.: Combining assessment tools for a comprehensive evaluation of computational thinking interventions. In: Kong, S.-C., Abelson, H. (eds.) Computational Thinking Education, pp. 79–98. Springer, Singapore (2019). https://doi.org/10.1007/978-981-13-6528-7_6

21. Rowe, E., et al.: Assessing implicit computational thinking in Zoombinis puzzle gameplay. Comput. Hum. Behav. **120**, 106707 (2021)

22. Statter, D., Armoni, M.: Teaching abstraction in computer science to 7th grade students. ACM Trans. Comput. Educ. **20**(1), 8–837 (2020)

23. Stupurienė, G., Jevsikova, T., Juškevičienė, A.: Solving ecological problems through physical computing to ensure gender balance in STEM education. Sustainability **14**(9), 4924 (2022)

24. Sun, L., Hu, L., Zhou, D., Yang, W.: Evaluation and developmental suggestions on undergraduates' computational thinking: a theoretical framework guided by Marzano's new taxonomy. Interact. Learn. Environ. (2022). https://doi.org/10.1080/10494820.2022.2042311

25. Tang, X., Yin, Y., Lin, Q., Hadad, R., Zhai, X.: Assessing computational thinking: a systematic review of empirical studies. Comput. Educ. **148**, 103798 (2020)

26. Tikva, C., Tambouris, E.: Mapping computational thinking through programming in K-12 education: a conceptual model based on a systematic literature Review. Comput. Educ. **162**, 104083 (2021)

27. Troiano G., et al.: Is my game OK Dr. Scratch? Exploring programming and computational thinking development via metrics in student-designed serious games for STEM. In: Proceedings of the 18th ACM International Conference on Interaction Design and Children Association for Computing Machinery, New York, NY, USA, pp. 208–219 (2019)

28. Wei, X., Lin, L., Meng, N., Tan, W., Kong, S.-C., Kinshuk.: The effectiveness of partial pair programming on elementary school students' computational thinking skills and self-efficacy. Comput. Educ. **160**, 104023 (2021)

"I Now Feel that this is Unfair" A Case Study on the Effects of Professional Development for Debugging in the K-12 Classroom

Tilman Michaeli[1]([⊠]) [ID] and Ralf Romeike[2]

[1] TUM School of Social Sciences and Technology, Computing Education Research Group, Technical University of Munich, Munich, Germany
`tilman.michaeli@tum.de`
[2] Computing Education Research Group, Freie Universität Berlin, Berlin, Germany
`ralf.romeike@fu-berlin.de`

Abstract. Finding and fixing errors is an essential skill in learning programming in the K-12 classroom. However, most of the time, debugging only plays a minor role in teachers' approaches to conveying programming - especially as they themselves rarely learned debugging explicitly and lack appropriate concepts and content. In consequence, students often struggle with finding and fixing errors on their own. Professional development allows for disseminating research findings and corresponding teaching materials to eventually influence the teaching practice. In this paper, we present a professional development workshop and its theoretical foundations, aiming at fostering teachers' professional competence with regards to debugging. We investigate changes in teaching practice and the teachers' beliefs in reaction to the PD using a case study approach. The results provide insights into impact and effects of professional development with regards to debugging in the classroom. Furthermore, our study contributes indications for designing professional development that fosters actual change in the classroom.

Keywords: Debugging · Professional development · Computing education · K-12 · Case study · Teaching practice

1 Introduction

Debugging can be considered a core problem in the K-12 classroom: novice programmers make more programming errors and, compared to experts, spend similar high amounts of time debugging [2]. Fixing errors is a significant obstacle to learning programming [20]. Helplessness and, in consequence, frustration when confronted with errors is a common phenomenon in the K12 classroom [26]. Accordingly, this is also a major challenge for teachers: They often rush from student to student, helping and trying to do justice to all of them as much as possible [24]. Moreover, teachers - just like professional software developers [27] - have often not learned debugging systematically themselves and only seldom explicitly teach debugging [24]. In school practice, learners are therefore often left alone with their errors and are consequently forced to acquire appropriate strategies and approaches on their own. Experience has shown that this is a challenge that is hardly manageable for a large part of students [6]. However, explicitly teaching debugging has the potential to foster students' debugging self-reliance [3,7,25].

As to why they don't convey debugging skills in their classroom, teachers report that there is a lack of time - both in the classroom and for preparing appropriate concepts and materials. At the same time, teachers claim a lack of existing concepts, best practices, or materials for debugging in the classroom. Furthermore, they report that debugging is not an explicit part of the curriculum and therefore often neglected in favor of content explicitly required [24].

A traditional way to transfer educational innovations and achieve a change in the teaching practice is professional development (PD) for in-service teachers. However, designing effective PD provides a particular challenge, as typically in limited time, teachers' content and pedagogical content knowledge as well as beliefs need to be addressed - while demonstrating direct applications and implications for their pedagogical practice [22]. Therefore, in this study, we designed a corresponding professional development workshop and investigate the effects on the participants' teaching practice in the form of a case study to gain insights on how to tackle the core problem of debugging in the classroom.

2   Theoretical Background

The aim of PD is to improve the quality of teaching and thus the students' learning processes [9]. Research shows that this is highly dependent on teachers' *professional competence* which is considered an acquirable disposition [19]. Therefore, expanding teachers' professional competence is central for PD. It is comprised of cognitive, motivational and personal components [4]. Cognitive components are typically distinguished into content knowledge (CK), pedagogical knowledge (PK) and pedagogical content knowledge (PCK) based upon [31]. Concerning motivational factors of professional competence, teachers' beliefs in particular are considered crucial.

In order to achieve a change in professional competence, according to constructivism, the teachers' existing experience has to be taken into account [32]. Debugging describes the process of finding and fixing errors and is dependent on the underlying error type [15]. Debugging skills differ from general programming skills [1] and include the application of a systematic process, different debugging strategies, heuristics and tools [21,24,30]. Teachers themselves typically have rarely learned debugging explicitly and therefore lack appropriate concepts and content for teaching in class. Furthermore, they differ in their personal debugging process: While some teachers apply predominantly (basic) strategies such as *print* debugging and have little experience with tools such as the *debugger*, others even teach such "advanced" methods to their students [24]. With regard to

PCK, teachers hardly use explicit units to teach debugging skills in their lessons. Predominantly, they try to give the students assistance in individual support. Despite acknowledging the problem coping with errors provides for students, teachers lack the means and/or beliefs to eventually address them further in their classroom [24].

There are some studies investigating explicitly teaching debugging in secondary or tertiary [3,5,7,8,17,25], but to our knowledge there is no research done on PD specifically focusing on debugging in the classroom. However, there are extensive findings regarding PD in [14,23,28,29] and especially beyond [9,11,19,22] computing education research. In the following, we discuss a selection of core design principles from literature that seem especially important for our context.

Research shows that for an actual change in school practice, it is crucial to *create a common problem awareness*. To this end, it is important to work together with teachers on the basis of their teaching practice, instead of merely providing "top-down" materials. It is essential that the teachers' own teaching experiences are taken into account. For the implementation of concepts of the workshop in teaching practice, it is helpful to provide materials that are as *concrete as possible*, making them more or less directly-usable in the classroom. This enables teachers to use the materials directly without much effort and thus reduces the entry barrier. However, teachers typically adapt these concepts and materials to their specific needs in a second iteration at the latest. Therefore, teachers should be actively encouraged to experiment and adapt the materials as a starting point for teaching. Further success factors for PD are an interplay between theoretical input and practical explorations phases that allow for *self-directed and active learning*. Due to the heterogeneity of teachers' knowledge, they should be enabled to try out new debugging strategies in a self-determined way, for example, according to their individual learning requirements. Additionally, *collaboration, the exchange of ideas and networking* with and between the participating teachers is essential for a sustainable transfer of the workshop content into teaching practice.

Those principles derived from research on PD provide hints for designing general PD. However, designing a successful workshop is highly dependent on the content and context [22]. Therefore, research in PD on debugging and its effects on the teaching practice is necessary.

3   A PD Workshop on Debugging

Based on the principles established in research on general PD as well as findings on teachers' professional competence with regards to debugging, we designed and conducted a weekend PD workshop (Friday afternoon to Sunday noon) in November 2019 with 16 participating CS high school teachers from various regions of Germany. The event took place in a conference hotel, so that the participants stayed overnight. With the workshop, we aimed at fostering teachers' professional competence regarding debugging (in particular concerning beliefs,

CK and PCK), for eventually influencing their teaching practice. For the classroom materials used in the workshop, we mainly build upon a teaching concept for including debugging in the classroom, which effectiveness we investigated previously [25]. Within this concept, different types of errors are distinguished and a systematic approach for fixing them is proposed. Furthermore, concrete methods for the classroom are suggested, such as adapted debugging tasks, ways to introduce various debugging strategies and tools (such as print-debugging or the debugger), or an error glossary to help with forming patterns and heuristics (similar to professionals' debugging logs [27]).

Before the workshop began, teachers were asked to reflect on their personal debugging process and how they had learned to debug in the form of "debugging biographies" (see e.g. [18]). The evaluation of these biographies revealed that the majority of teachers never learned debugging systematically, but had mostly acquired the relevant skills on their own. Furthermore, it showed a large amount of heterogeneity with respect to the scope of the debugging strategies and tools used and also concerning the general programming experience outside the classroom.

Day 1: Awareness and self-reflection. At the beginning of the workshop, debugging was characterized as a core problem of teaching practice. To this end, the teachers first reflected how "typical" students proceed with debugging in their lessons and what problems they have using the *persona method* from Design Thinking [10] in groups (see Fig. 1). In doing so, they switched into the perspective of the learners to reflect on typical problems (a central component of *PCK*). In a second step, the groups exchanged the personas among themselves and described how they would support the respective student in their current teaching. This way, teachers should both share existing best practices and reflect on where they see a need to expand their teaching.

Day 2: Introduction and exploration of CK and PCK on debugging. At first, a professional software developer reported from his professional experience on how debugging works in industry, what significance it has and how developers "learn" debugging. In the next phase, the debugging process was formally introduced and corresponding debugging skills were systematized (*CK*). Furthermore, approaches to address respective required skills in the classroom were demonstrated (*PCK*). Afterward, the teachers worked through materials for classroom use. In doing so, they acquired corresponding *CK*, for example by working on materials for the introduction of debugging strategies with which they had little or no experience (such as the debugger). Furthermore, they directly tried out methods for teaching relevant skills and thus acquired according *PCK*. All ideas, suggestions, questions, and comments that arose during this phase were collected and then discussed in plenary.

Day 3: Transfer into their classroom. On the last day of the workshop, teachers were given the opportunity to transfer the concepts to their own teaching materials. Based on the personas, each teacher created a concrete plan of which ideas

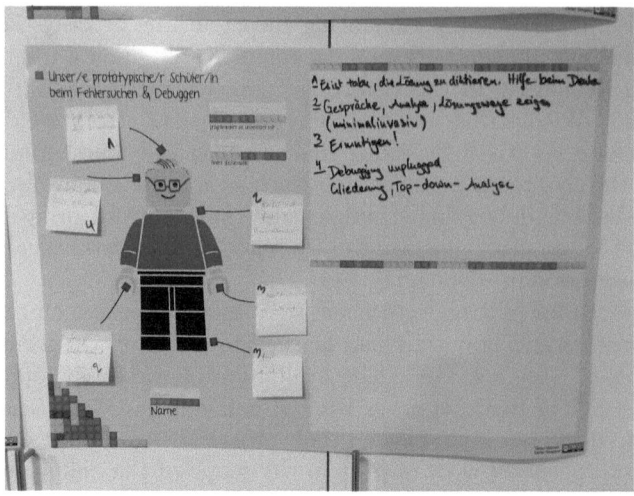

Fig. 1. Persona from the workshop

they wanted to take from the workshop and use in the classroom. These plans were then presented and discussed in groups.

Concluding reflection in the workshop. The design of the workshop was generally viewed positively by the teachers. Among other things, they highlighted the high proportion of active learning, the expert's input and the intensive networking among the teachers. The opportunity to try out concrete materials for the classroom was also positively evaluated. A clear consensus for the increased integration of the topic of debugging into teaching became apparent.

4   Methodology

This study aims to investigate the influence of the PD workshop on the actual teaching practice. Therefore, we address the following research question. **RQ:** How does the teaching of the participants involved in the PD workshop change with regards to debugging in the classroom?

There are many ways to evaluate the success of a PD workshop. Since we are explicitly interested in the transfer to teaching practice, we aimed to investigate the actual change in the teachers' classrooms – as opposed to an examination of the change in self-efficacy expectations or professional competence directly at the end of the training [22]. However, shortly after the workshop in November 2019, school practice was largely restricted for the survey period due to the Covid-19 pandemic. Given the severe additional challenges and demands for teachers – not just limited to remote teaching periods –, an in-detail comparative analysis does not appear to be expedient. Furthermore, it is to be expected that the time available and willingness to implement new concepts was severely limited.

Therefore, under these special conditions, a qualitative case study methodology was chosen to investigate the research question. This allows for a precise analysis of cases for teachers reporting to have adapted their teaching in consequence of the workshop, despite or even before the pandemic effects on schools. Furthermore, we can examine the characteristics of the individual cases and the respective circumstance in detail to comprehensively map the changes in the teaching practice [33].

Data Collection and Case Selection. Towards the end of the school year, semi-structured interviews were conducted online with the participants of the workshop. In the process, *critical cases* [33] were selected, which allows to check the connection between the workshop and changes in the teaching practices. Despite the Covid-19 pandemic, two of the teachers contacted had already integrated corresponding content from the PD workshop into their lessons to a significant extent before the school closures began, or had implemented it in remote teaching despite the corresponding challenges. Both were well-experienced CS high school teachers from different German states (with different curricula and ways of anchoring CS as a subject). Within the semi-structured interviews, the teachers were asked whether and how their teaching changed in consequence to the PD, in particular with regards to their awareness of and reaction to students' problems, as well as whether they tried out the methods proposed – and if so, which experiences they made.

Data Analysis. The interviews were first transcribed and evaluated in a case-by-case analysis (*within-case analysis*) to develop a deeper understanding of the respective changes [12]. Subsequently, central *cross-case* characteristics were identified.

5   Results

In the following section, the two cases are described in detail, in order to then identify common characteristics afterwards. All quotes have been translated into English by the authors with minimal adjustments to improve comprehensibility.

5.1   Teacher I

Teacher I reports that shortly after the end of the workshop, he already implemented the first concepts regarding a systematic debugging approach with year ten students:

> *As soon as that was possible, I tried it. [...] We categorized [errors] by type and also discussed how to cope with them. There were actually two lessons where that was the topic. The students had a given programming project, and there were errors built-in, and they had to solve different tasks, [...] from a semicolon that was missing to semantic errors that appeared at the end, even though the program was running perfectly.*

He applied debugging tasks and let the students categorize different error types and emphasized that a different approach was necessary for different kinds of errors. In general, categorizing errors with the students represents one of the central changes in the teaching. The teacher was also able to give his assessment of success and impact on the students before the Covid-19-based distance learning:

How helpful this is in the long run is difficult to confirm or disprove at the moment. For the weaker students it has definitely been helpful. They have at least gratefully accepted this categorization. And they [...] then wrote an error glossary, and I think they pulled it out again when they had errors that they didn't know about at first and had to remember.

With the joint error collection in the form of a "glossary", another idea of the workshop was adapted and successfully implemented. The teacher's experience indicates that "weaker" students in particular have actually benefited from the support. In general, the teacher emphasizes that in the future he wants to introduce the handling of errors at the beginning of the programming lessons.

In addition, the teacher has transferred debugging to the topic of spreadsheets and also used "debugging tasks" here:

So it was interesting to see what kind of errors there were. They often have problems there as well. You have to run around because too many hash marks are displayed, which simply indicates that the columns are too narrow. Typical errors are division by zero and so on. And the error messages are quite cryptic.

Dealing with errors is an overarching theme for the teacher after the workshop and is not limited to programming. This emphasizes the general educational importance of debugging beyond programming. As a consequence of the workshop, the teacher focuses especially on the error culture in class:

So another aspect we covered in this context in the lesson before Christmas, but that I have also done occasionally in recent years, is to discuss famous software bugs. Making errors, Ariane and Mars Lunar Voyager and so on, all that stuff. These are exciting stories, which also show the students that even on a large scale, mistakes are made and that errors just happen.

After reflecting his teaching, he assumes that there is a connection between the opening of tasks and the perception and handling of errors:

The tasks are often such that the students have to solve a problem very specifically. The task is clearly defined. The students have to solve it and then stumble somewhere into these mistakes they then make and are then disappointed because it doesn't work out right away. [...] This means that in every lesson you get to the point when the students make mistakes: "Now I have made a mistake. I didn't solve the problem the way the teacher intended." And this should actually be different in the whole programming class: that I work more creatively and simply give the students more freedom in the exercises.

In the future, he therefore wants to use more open assignments in the initial lessons to test what influence this has on how students deal with their errors. This can be described as an *productive-failure* approach [16]. Overall, the importance of the topic for teaching has thus increased and is also multiplied in his training of teachers:

From practice I can still say that I am now also addressing this issue with teacher trainees. With them, I will cover this as a explicit topic in the next few weeks, because currently we were hindered by Covid. But now, until the end of the school year, it is also my goal that we explore approaches [for debugging in class].

In addition, the teacher also reports that his own approach to debugging has evolved in the context of programming projects in a newly-learned programming language:

There I learned something I never did and never needed in Java, which we discussed at the workshop, namely print debugging. I need it once or twice for some Java problems, but with the Python problems it was so massive. [...] That means, you always have to look, what is input, what is output, what is the current state of the variables? And I did an incredible amount of print debugging. Which I have never actually done before.

In summary, the teacher reports a significant change in his classroom as a result of the PD workshop. Coping with errors is now perceived as a content that spans all topics, students are supported particularly well by systematizing and collecting various error types, and the teacher tries to create a positive error culture in the classroom.

5.2   Teacher II

The second teacher also reports that he has integrated debugging more prominently in his teaching as a result of the workshop. Accordingly, he introduced two debugging strategies for the first time during distance learning. First, a "debug class", which logs the calls of the methods of a given project:

I introduced a debug class, because I am looking for errors myself and with the help of this debug class, traces were created and these traces should be displayed in sequence diagrams to use the aspect of modeling [...], but also to make basic mechanisms clear. That means we did not learn from the error, but from the trace.

Thus, this debugging strategy was initially not introduced in the context of error correction, but rather in the context of learning about modeling, and only then should it be used for debugging. In addition, the debugger was introduced as an optional task in analogy to the approach of the teaching concept presented in the workshop, using a project created by the teacher. The teacher emphasized that these two approaches came from the workshop, since debugging is not an explicit topic in the curriculum.

He hopes that this change will give the students more autonomy in debugging, which will also enable them to carry out other kinds of teaching projects – for which he was unable to gain experience because of distance learning:

This is the innovation in my teaching or in my approach. That was not the focus in the past. The point was to keep the project so small that I actually assume that it would be flawless.

In conclusion, the teacher summarizes the changes in his teaching practice, especially concerning debugging:

What you have done, from my point of view, is you have influenced the beliefs of our colleagues. In consequence, I see the importance better than before now. I actually think that it is not quite fair to always reproach the student for not finding the errors when I have not even shown him what I normally use. I now feel that this is unfair. That's why the minimum I have to show him is what I usually use myself. I usually use tracing, so that's what I teach.

This shows a changed view on his teaching, which emphasizes that without the teaching of adequate strategies, students are not able to deal with errors on their own. The workshop has thus contributed to changing the teacher's beliefs concerning the importance of debugging for programming lessons.

In summary, the importance of debugging in class – although not anchored in the curriculum – has increased significantly for this teacher. Based on his own debugging approach, he now systematically introduces debugging strategies to increase the students' independence.

6   Discussion

Comparing the two cases concerning the change in teaching, the first thing to be noted is the increased **value** of debugging. The importance of debugging in the classroom – even beyond programming – is reflected in (increased) time spent on explicitly conveying debugging skills by the teachers.

Neither teacher has directly adopted the materials from the workshop, but both have **adapted** them for their own needs (in line with literature [13]). In doing so, they have set different priorities: For example, teacher I focused particularly on the aspect of error culture and the categorization of different types of errors, while teacher II primarily conveyed his personal debugging approach and strategies to students. In both cases, this results in a **extension** of the workshop contents: For example, teacher I transfers the concepts of the PD to the teaching of spreadsheets, while teacher II combined the introduction of appropriate tracing strategies with teaching modeling.

As a consequence of the workshop, both teachers also **reflect** their teaching practice with regard to dealing with errors: They would like to try out more open tasks in the future. Teacher II hopes improving students debugging skills allows him to use such open formats. Teacher I suspects a connection between more open tasks and the perception and handling of errors according to a *productive-failure* approach, which he would like to explore in the future.

The basis for these changes in teaching practice is the **expansion of professional competence** in consequence of the workshop: On the one hand, the teachers have acquired corresponding *content knowledge*, which even influenced their personal debugging process. On the other hand, the teachers learned about different ways to teach debugging skills (*pedagogical content knowledge*). At the same time, the *beliefs* of the teachers have changed towards the explicit teaching of debugging. In both cases, this had hardly played a role in their teaching practice before.

Limitations. Due to the Covid-19-related teaching situation and the corresponding challenges in the classroom, many participants of the workshop reported that they had not yet integrated debugging content into their lessons as planned. A broader evaluation of the PD regarding the transfer into the teaching practice was not possible. However, the cases of teachers who had already implemented debugging in their classes before or despite Covid19-circumstances allow for deep insights into the changes in their teaching and thus implications about success factors of the workshop.

7   Conclusion

In this paper, we investigated the effects a PD workshop on debugging in the K-12 classroom has on the teaching practice of attending teachers. For this purpose, a three-day workshop was designed based on general research findings regarding PD, as well as research on teachers' professional competence with regards to debugging in the classroom. Given the pandemic situation, we conducted a case study to analyze effects on participants' teaching. In consequence, our results provide deep insights into two teachers' change in practice.

The analysis revealed that the workshop had the intended effect for the two cases. This was particularly evident in the increased importance of the topic in teaching practice. In most cases, the materials of the workshop were the starting point for individual adaptation according to personal needs and different foci. In addition, the awareness on the significance of errors for the learning process introduced in the workshop sparked further ideas for extensions, such as transferring and combining the concepts with other topics.

Furthermore, the results provide indications for designing professional development that foster actual change in the classroom. To this end, our results suggest that convincing the teachers of the importance of the topic (teaching debugging in the classroom) is crucial: even if not explicitly required in the curricula, debugging is an essential part of the programming process. Fostering students' debugging skills even offers teachers the potential to improve their own teaching practice further, such as by including more open exercises. Furthermore, the teachers' reports indicate that it is precisely the reflection of their own teaching practice that has contributed to the creation of an awareness of the problem and was thus essential for changing teachers' beliefs towards debugging. Actual change in practice was supported by the concrete materials as possible starting points for teaching debugging skills: The teachers adapted the ideas of the workshop to their personal needs, expanded them and experimented with them – even beyond the programming classes.

References

1. Ahmadzadeh, M., Elliman, D., Higgins, C.: An analysis of patterns of debugging among novice computer science students. In: Proceedings of the 10th Annual SIGCSE Conference on Innovation and Technology in Computer Science Education, pp. 84–88. ACM, NY (2005)
2. Allwood, C.M., Björhag, C.G.: Novices' debugging when programming in Pascal. Int. J. Man Mach. Stud. **33**(6), 707–724 (1990)
3. Allwood, C.M., Björhag, C.G.: Training of Pascal novices' error handling ability. Acta Physiol. **78**(1–3), 137–150 (1991)
4. Blömeke, S., Felbrich, A., Müller, C., Kaiser, G., Lehmann, R.: Effectiveness of teacher education. ZDM **40**(5), 719–734 (2008)
5. Böttcher, A., Thurner, V., Schlierkamp, K., Zehetmeier, D.: Debugging students' debugging process. In: 2016 IEEE Frontiers in Education Conference (FIE), pp. 1–7. IEEE, Erie, PA (2016)
6. Carver, S., Klahr, D.: Assessing children's logo debugging skills with a formal model. J. Educ. Comput. Res. **2**(4), 487–525 (1986)
7. Carver, S., Risinger, S.: Improving children's debugging skills. In: Empirical Studies of Programmers: Second Workshop, pp. 147–171. Ablex Publishing Corp. (1987)
8. Chmiel, R., Loui, M.C.: Debugging: from novice to expert. ACM SIGCSE Bull. **36**(1), 17–21 (2004)
9. Clarke, D., Hollingsworth, H.: Elaborating a model of teacher professional growth. Teach. Teach. Educ. **18**(8), 947–967 (2002)
10. Dahiya, A., Kumar, J.: How empathizing with persona helps in design thinking: an experimental study with novice designers. In: IADIS International Conference Interfaces and Human Computer Interaction (2018)
11. Darling-Hammond, L., Hyler, M., Gardner, M.: Effective Teacher Professional Development. Learning Policy Institute, Palo Alto, CA (2017)
12. Eisenhardt, K.M.: Building theories from case study research. Acad. Manage. Rev. **14**(4), 532–550 (1989)
13. Farmer, J., Gerretson, H., Lassak, M.: What teachers take from professional development: cases and implications. J. Math. Teacher Educ. **6**(4), 331–360 (2003)
14. Goode, J., Margolis, J., Chapman, G.: Curriculum is not enough: the educational theory and research foundation of the exploring computer science professional development model. In: Proceedings of the 45th ACM Technical Symposium on Computer Science Education, pp. 493–498 (2014)

15. ISO, I.: IEEE, systems and software engineering-vocabulary. IEEE Computer Society, Piscataway, NJ 8, 9 (2010)
16. Kapur, M.: Productive failure. Cogn. Instr. **26**(3), 379–424 (2008)
17. Katz, I., Anderson, J.: Debugging: an analysis of bug-location strategies. Hum. Comput. Interact. **3**(4), 351–399 (1987)
18. Knobelsdorf, M., Schulte, C.: Computer biographies-a biographical research perspective on computer usage and attitudes toward informatics. In: Proceedings of the Koli Calling 2005, pp. 139–142. ACM, NY (2005)
19. Kunter, M., Kleickmann, T., Klusmann, U., Richter, D.: The development of teachers' professional competence. In: Kunter, M., Baumert, J., Blum, W., Klusmann, U., Krauss, S., Neubrand, M. (eds) Cognitive Activation in the Mathematics Classroom and Professional Competence of Teachers. Mathematics Teacher Education, vol. 8, pp. 63–77. Springer, Boston (2013). https://doi.org/10.1007/978-1-4614-5149-5_4
20. Lahtinen, E., Ala-Mutka, K., Järvinen, H.M.: A study of the difficulties of novice programmers. ACM SIGCSE Bull. **37**(3), 14–18 (2005)
21. Li, C., Chan, E., Denny, P., Luxton-Reilly, A., Tempero, E.: Towards a framework for teaching debugging. In: Proceedings of the Twenty-First Australasian Computing Education Conference, pp. 79–86. ACM, NY (2019)
22. Lipowsky, F., Rzejak, D.: Key features of effective professional development programmes for teachers. Ricercazione **7**(2), 27–51 (2015)
23. Menekse, M.: Computer science teacher professional development in the united states: a review of studies published between 2004 and 2014. Comput. Sci. Educ. **25**(4), 325–350 (2015)
24. Michaeli, T., Romeike, R.: Current status and perspectives of debugging in the K12 classroom: a qualitative study. In: 2019 IEEE Global Engineering Education Conference (EDUCON), pp. 1030–1038. IEEE, Dubai (2019)
25. Michaeli, T., Romeike, R.: Improving debugging skills in the classroom: the effects of teaching a systematic debugging process. In: Proceedings of the 14th Workshop in Primary and Secondary Computing Education, pp. 1–7. ACM, NY (2019)
26. Perkins, D.N., Hancock, C., Hobbs, R., Martin, F., Simmons, R.: Conditions of learning in novice programmers. J. Educ. Comput. Res. **2**(1), 37–55 (1986)
27. Perscheid, M., Siegmund, B., Taeumel, M., Hirschfeld, R.: Studying the advancement in debugging practice of professional software developers. Softw. Qual. J. **25**(1), 83–110 (2016). https://doi.org/10.1007/s11219-015-9294-2
28. Qian, Y., Hambrusch, S., Yadav, A., Gretter, S.: Who needs what: recommendations for designing effective online professional development for computer science teachers. J. Res. Technol. Educ. **50**(2), 164–181 (2018)
29. Ravitz, J., Stephenson, C., Parker, K., Blazevski, J.: Early lessons from evaluation of computer science teacher professional development in Google's CS4HS program. ACM Trans. Comput. Educ. **17**(4), 1–16 (2017)
30. Rich, K.M., Strickland, C., Binkowski, T.A., Franklin, D.: A K-8 debugging learning trajectory derived from research literature. In: Proceedings of the 50th ACM Technical Symposium on Computer Science Education, pp. 745–751. ACM, NY (2019)
31. Shulman, L.: Knowledge and teaching: foundations of the new reform. Harv. Educ. Rev. **57**(1), 1–23 (1987)
32. van Dijk, E.M., Kattmann, U.: A research model for the study of science teachers' PCK and improving teacher education. Teach. Teach. Educ. **23**(6), 885–897 (2007)
33. Yin, R.: Case Study Research: Design and Methods, 3rd edn. Sage, London (2003)

Robotics-Enhanced Natural Science in Primary Schools

Bence Gaál[✉] [iD]

Faculty of Informatics, Eötvös Loránd University, Pázmány P. sny 1/C, Budapest 1117, Hungary
gaalbence@inf.elte.hu

Abstract. The challenges of the 21st century and the modern age require people to have knowledge of the natural sciences and information technology [1, 2]. Ideally, these two disciplines should be combined in everyday life inside classrooms. However, this is often not the case, and students often only have access to such activities outside the classroom (e.g.: in workshops). As part of my research, I develop good practices that allow us to integrate projects from computer science (digital culture) classes into the natural science classes.

In my article I would like to present the first experiences of the practical implementation of my doctoral research, in which I am implementing natural science lessons with 5th grade children, where robotics is used as an illustrative, and modelling tool for different topics. This method has provided an opportunity to link information technology (IT) and natural science in primary schools, and to allow interoperability between the subjects in the framework of different projects.

My class of 22 students and I are using robotics as a visual aid in several places per topic. The specificity of my model is that students make these teaching tools themselves during IT lessons. In the process, children also acquire a basic knowledge of programming. The projects are carried out in groups, which allows them to develop several soft skills during these IT lessons, this adaptability and development are also very important skills in this day and age [3]. The natural science lessons are also carried out in a manner that provides new and exciting exemplification for all students, which is designed to increase and maintain motivation. In this article I would like to describe the themes of the lessons and the tools used. I will then describe children's results and their views on this way of learning, analysed through questionnaires and interviews.

Keywords: STEM · Robotics · Natural science

1 Introduction

Today, STEM subjects have a high priority in the labour market and unemployment rates in this field are below the EU average [4]. From the curricula that are built around and focuses on STEM, the good ones are those that provide opportunities for integrated education [5]. In Hungary, the National Curriculum and the Framework Curriculum based on it provide this possibility, but do not give teachers sufficient support or help in the practical application [6]. Therefore, it is important to develop a methodology

A. Bollin and G. Futschek (Eds.): ISSEP 2022, LNCS 13488, pp. 90–100, 2022.
https://doi.org/10.1007/978-3-031-15851-3_8

that would stimulate children's motivation in these areas and allow for a high level of integration in natural science and technology, taking European educational trends into consideration. This would be necessary also because Hungary is below the EU average in the number of BA and MSc degrees in STEM subjects [7] and because the number of students entering higher education in STEM fields is decreasing [8].

Therefore, I would like to present a method that integrates robotics to a high degree in the natural sciences and uses technology as an illustrative tool in everyday lessons, but not in a ready-made form, rather prepared by the students. In this way, their skills in STEM, or more recently STEAM, are strengthened, as their creativity [9] and other soft skills are developed through design and implementation.

2   Presentation of Lessons and Tools Used

The lessons were implemented in normal classroom conditions in the upper, 5th grade of public education. It was essential that I taught both the natural science and IT subjects to the students. The group consisted of 22 students (11 girls, 11 boys). Two important factors were essential for the implementation. The first was the possibility of cross-curricular transfer. As I am also the class teacher, I had the flexibility to use the group's lessons. The other factor is having the right IT tools. At the moment, I think that this method can be achieved if one teacher teaches the two subjects. Unfortunately, this kind of cooperation between teachers is not really present in the Hungarian education. In the structures of the lessons, I will explain in more details why a very high level of cooperation would be necessary during these lessons.

2.1   The Tools Used

After reviewing the available IT devices on the market, and considering the feasibility, I decided to use the BBC micro:bit V2 board (Fig. 1). Several studies in the UK have demonstrated the effectiveness of these tools, highlighting that they can help girls to become more involved in IT and that a high proportion of girls would choose this career [10].

The hardware design of the device also influenced the decision, as the new version of the micro:bit chip has a speaker and a faster processor, and its energy-saving mode allows for longer use and also demonstrating the importance of sustainability for the young generation [11]. Finally, the price/performance ratio and the compactness of the device were considered, allowing the devices to be moved from one classroom to another. The tool was programmed in a block environment by the children.

However, given the specificities of natural science, it was necessary to expand the number of sensors[1] installed, since many of the processes that take place in nature are due to some other effects. For the extension we used the KS0361 (KS0365) keyestudio 37 in 1 Starter Kit for BBC micro:bit (Fig. 2).

---

[1] The basic device has the following sensors: accelerometer, compass, light sensor, thermometer, microphone.

Fig. 1. The new BBC micro:bit [11]

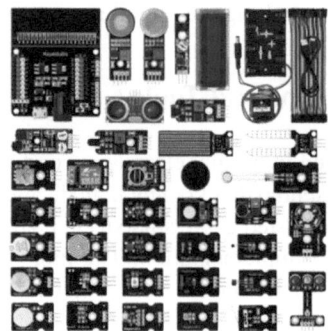

Fig. 2. Keyestudio 37 in 1 Starter Kit for BBC micro:bit [12]

2.2 Structure of the Lessons

The first step in delivering robotics-enhanced natural science lessons was to change the structure of the IT (now called digital culture) lessons. During the first semester, this was easier because the focus was on programming by default.

Children were introduced to microbits in a special way, as in some cases during the learning process we made tools with them in the digital culture lesson, which act as visual aids in the natural science lesson, as tools for independent models or experiments. And the integration of the use of these tools is and has been ongoing as the science curriculum has progressed throughout the years.

Considering this, every natural science lesson where we have used or will use the microbit should be preceded by a digital culture lesson where children could create and program the tools that we would later use in our learning. So, the subjects had to be in perfect sync with each other. If it had been implemented with more teachers, it would have required daily communication, full transparency and interoperability between the two subjects. I believe that this would only be possible in a few exceptional cases in our country. The situation is further complicated by the implementation of the digital culture curriculum at a time when the main focus is not on programming but on other subjects. In such cases, a flipped classroom approach was adopted, with a focus on independent task solving, and micro:bit emerged as a recurring curriculum for 1–1 lessons. The advantage of this is that the children's programming knowledge is kept up to date. The disadvantage may be that it requires more effort from the teacher's point of view to hold possible consultations and manage the class.

2.3 Presentation of Specific Topics Processed Using Robotics

During the first semester, two questionnaire surveys and one interview session were carried out. At the time of writing, three additional curricula have been integrated, but the results are still to be evaluated and the semester grades will provide a basis for comparison.

In addition to these, there were two smaller installations of micro:bit, but these were not surveyed separately. In one case, a purring kitten was implemented at the touch of the touch sensor, while in the other case I demonstrated the principle of the magnetic doors using a magnetic sensor, illustrated with an LED bulb. These demonstrations were not assessed separately because they were not created by the students but were presented as demonstrations used in a traditional way during lessons, as a simple experiment would be.

Body Structure of Plants - The Germination Process. In this project, we created a simulation to show students how the germination process takes place in nature. The existence of the conditions for the initiation of germination was monitored using a thermometer (P0) and a moisture sensor (P1). If all the external factors were present, the micro:bit display showed a plant emerged from the seed (Fig. 3).

The exercise is suitable for demonstrating simple condition checking and logical relations. Development possibility: to show the wilting process, the plant withers if it does not get enough light.

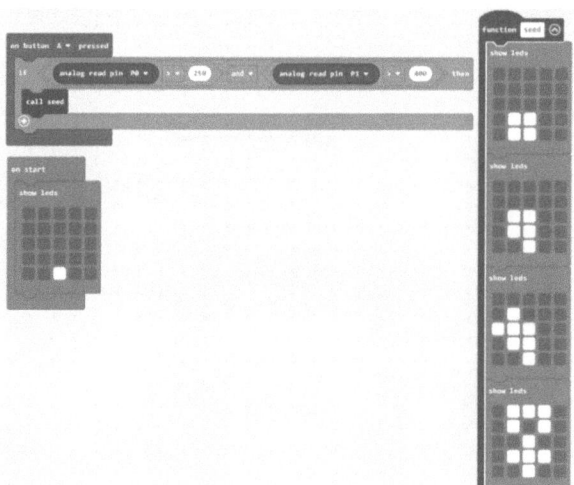

Fig. 3. The code of the germination project[2]

Animals Body Structure - the Honeybee. During the project, we needed two micro:bits that communicated with each other via a radio link. One device acted as

---

[2] https://makecode.microbit.org/_aVJVcWaEKaMM.

the flower and the other as the bee. The micro:bit acting as the bee made a buzzing sound when it moved, while the display showed an animation of a flying bee (Fig. 4). The other device initially displayed a flower on a tree, which would turn into a crop if the bee spent enough time near the flower to pollinate it (Fig. 5).

The exercise is suitable for demonstrating the principles of sending different radio packets, programming switches, and using variable handling.

A further link with mathematics was the introduction of the concept of absolute value, which was needed to handle the displacement of the micro:bit, since the acceleration strength alone did not provide a solution due to the effect of gravity.

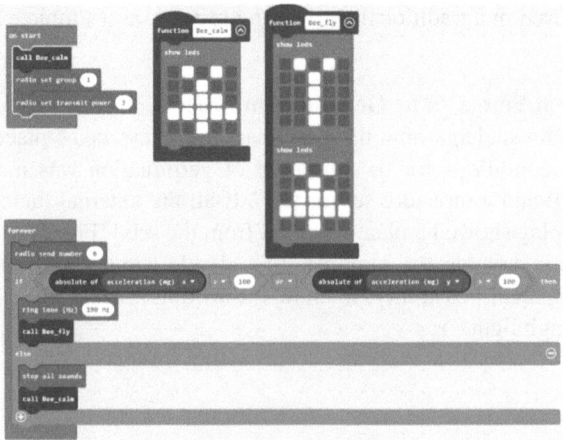

Fig. 4. The code of the bee[3]

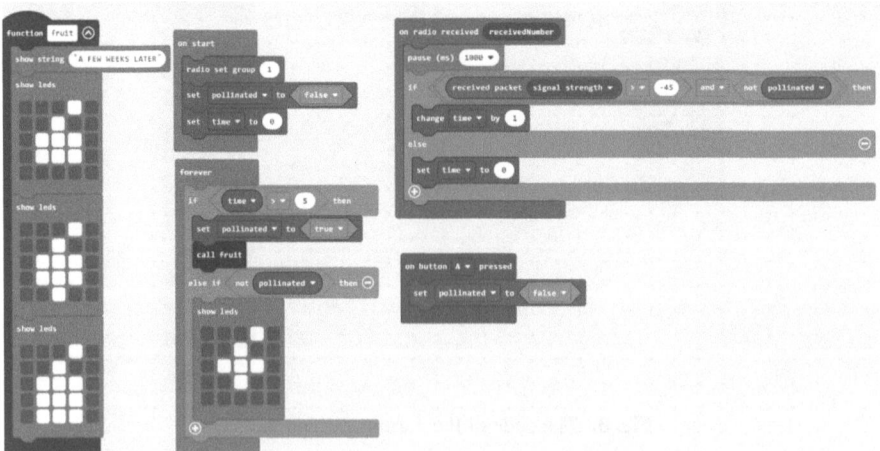

Fig. 5. The code of the plant[4]

---

[3] https://makecode.microbit.org/_Yo3YcKi1MMPf.

[4] https://makecode.microbit.org/_hFk4HACXqcRr.

Materials and Their Properties - Fire Alarm. In the next project with fire protection, we created a fire alarm with an LCD display that sends a text message to the user while a siren sounds if the flame sensor detects a fire.

The exercise is a good way to introduce cycles and to demonstrate how to connect the device to an external display, and to link the concept of the frequency of sound to physics. It can also be used to introduce the concept of multithreaded programming (Fig. 6).

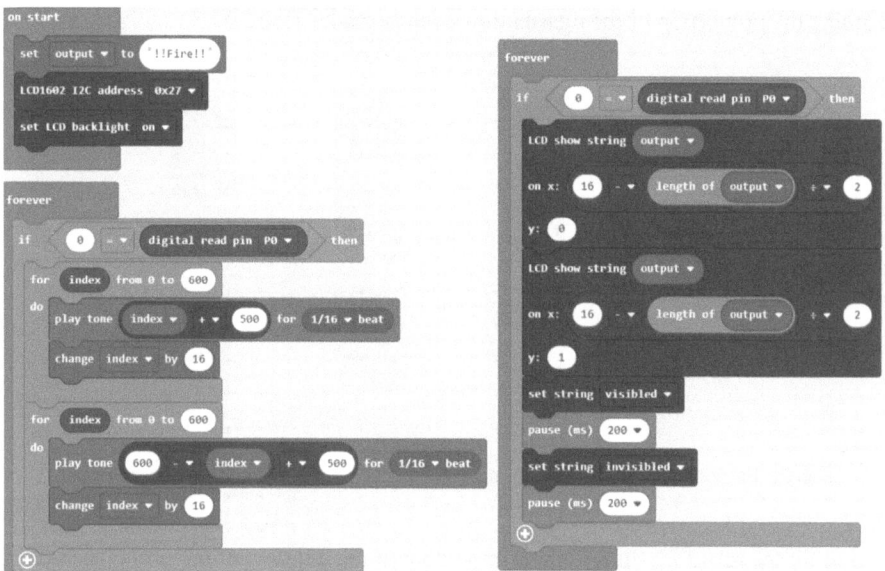

Fig. 6. The code of the flame detector

3   Students' Results Compared to Previous years

My students' grades are shown in the graph below in comparison to the results of the previous year group and the current 5th grade students.

It should be noted that during the school year, the grading options from teacher to teacher and the content of the end-of-term tests may vary, they are not standardised, but all classes follow the framework curriculum and use a uniform book on which teachers base their tests. Although the emphasis may differ in some places, but the topics and the outcome objectives are in any case the same. Accordingly, I have used the end-of-semester average rather than the total marks for my comparison (Table 1).

Looking at the trend of the semester averages, the best result is obtained by group 5.a, where the experiment is conducted. In terms of score, they scored 0.25 higher than any group in the last two years during the semester.

Table 1. Semester averages and class sizes of the examined classes

	5.a(2021–22)	5.b(2021–22)	5.a(2020–21)	5.b(2020–21)
Averages	4,64	4,20	4,39	4,11
Headcount	22	15	23	27

However, the groups differ greatly in terms of headcount, so it may be interesting to examine the proportion of the distribution of each grade[5] (Fig. 7).

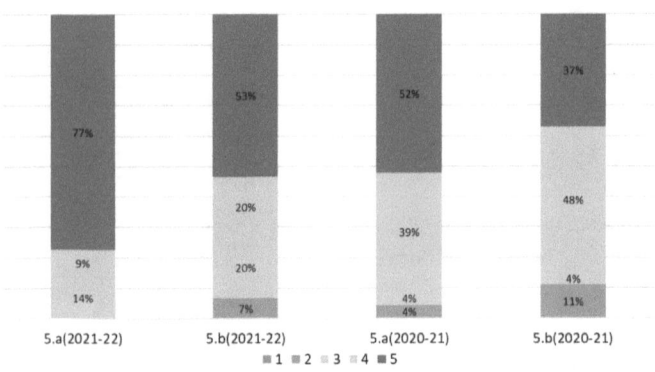

Fig. 7. Proportional distribution of natural science marks by class for the semester

I think the result is still telling, as more than three quarters of the group scored the highest marks, and no one scored a 2 compared to the other groups.

4   Feedback from Students

4.1   Feedback Through the Interview

I measured the students' impressions after the first session through an interview. In this context, I asked them if they had ever done programming before, how did they like this type of nature class, if they would like more, and what their feelings were and are before and after the session.

Of the participants, 5 had prior knowledge of programming. The session was not considered difficult by the majority, although there were 4 people who considered it moderately difficult. In the session where the tool was demonstrated, apart from one person who was not present for the demonstration, everyone gave positive feedback on how they felt about it. Unanimously, everyone wanted more classes like this. In this regard, 5 of the children said that the integration of robotics made it easier to understand the natural science material, 5 found the presentation of the processes interesting and

---

[5] In Hungary, this is on a scale of 1 to 5, with 5 being the best available.

good, and the rest gave the feedback that they simply liked it or found it exciting. From these results it seems that for everyone this was a positive experience, which could even have a positive impact on learning and the learning process. When asked how they felt before the lesson, most of the answers showed excitement and that they could not quite imagine how it could be done.

I also discussed the collaborative way of working during the interviews. All the respondents described positively what it was like to work in a self-organising team. One team had a problem with a team member with a confirmed attention deficit. The team was very patient with him, and this was reflected in his perspective as he was having a good time. Half of the respondents highlighted the fact that they could help each other and rely on each other. This confirms that the positive effects of this way of working includes building confidence, self-esteem, and a supportive environment [13].

None of the students would change the implementation of making, programming, and testing together in computer science class, and then using the tools as a demonstration tool in science class.

4.2 Results of the Two Questionnaire Surveys

In the following, I would like to review the results of the questionnaires completed after the two sessions. These questionnaires were conducted after the models were made, used, and presented. The questionnaires were completed digitally. Anonymity was not expected, as I developed the lessons based on these questionnaires in the meantime, however, giving their names was not mandatory. The relevant questions asked during the research were:

1. *How interesting does micro:bit make natural science lessons?*
2. *How much does micro:bit used in science class help you understand the given part of the curriculum?*
3. *How good do you think it is that I present certain parts of science with the help of robots?*
4. *Are you looking forward to the next time we use robots in a natural science class?*
5. *How exciting do you find these natural science lessons?*
6. *How often should micro:bits be used in natural science lessons?*

For **question 1**, there was one case of negative feedback on the first topic. In both cases, more than three quarters of the students said that the tool made the lessons interesting. The student who gives negative feedback is the finest student in the class. The background of the negative feedback was the new kind of tasks. In the process, it was often necessary to invent the solutions with their own ideas, and it was not possible to prepare for them at home, so the source of the problem was the fear of uncertainty in his case. In my opinion, one of the greatest problems of the Hungarian education system is that it gives little room for creative thinking and it puts more emphasis on lexical knowledge (Fig. 8).

For **question 2**, the numbers are more evenly distributed. Negative feedback was also received from one student, who in this case is the same respondent as the one who gave negative feedback earlier. At the end of the questionnaire, he commented that natural

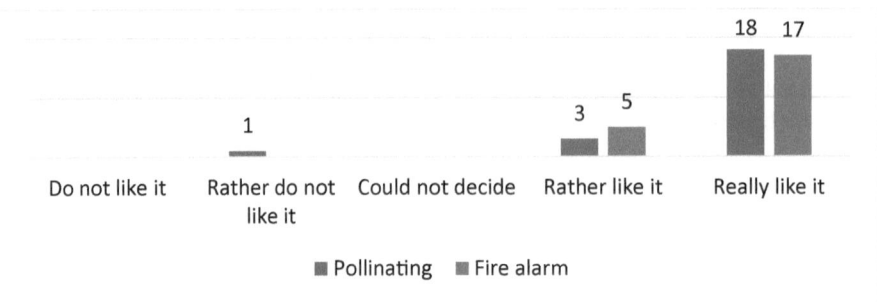

Fig. 8. Question 1: How interesting does micro:bit make natural science lessons?

science is interesting enough without robots. In my opinion, it will not be possible to talk about trends for this question in the future, since the difficulties of the processed material are not always the same, in this case the pollination process is much more complicated than the operation of a fire alarm (Fig. 9).

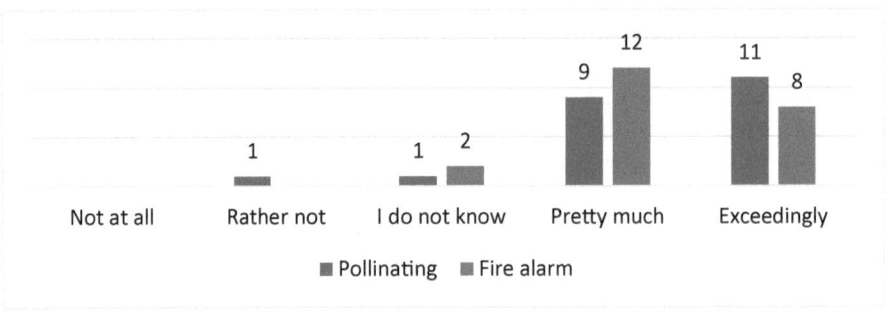

Fig. 9. Question 2: How much does micro:bit used in science class help you understand the given part of the curriculum?

For **question 3**, everyone gave positive feedback in both cases. The first session was marked in a 18–4 ratio and the second session was marked in a 20–2 ratio with the very and the quite options.

In **question 4**, 1 person indicated that they did not want to do more of the pollination project, 1 that they did not know and 20 that they would like to do more. For fire alarms, 21 people would like more and 1 did not know.

In response to **question 5**, 1 person also indicated that they were less looking forward to the next opportunity. 4 students were mostly looking forward to the next occasion and 17 were very much looking forward to the next occasion. After the fire alarm project, 4 people are "mostly" looking forward to the next robotics-enhanced natural science lesson and 18 are very much looking forward to the next one (Fig. 10).

According to the answers to **question 6**, after the pollination lesson, 17 people said that we should use the tools as often as possible. This number was 19 after the fire alarm. The monthly option was selected by 4–2 people and every two to three months option was selected by 1–1 people.

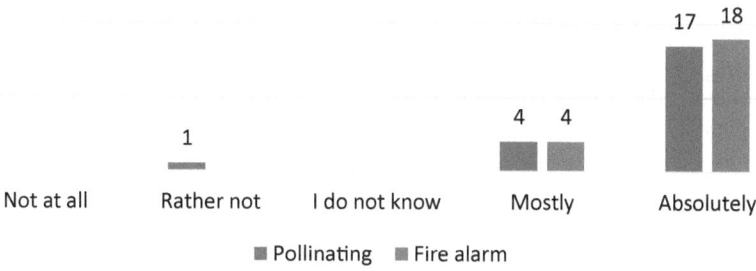

Fig. 10. Question 5: How exciting do you find these natural science lessons?

5 Summary

Overall, the questionnaires show that, looking at the average of the responses, we can see the beginning of an upward trend, with one exception. The aim of the sessions is to keep children's motivation for natural science and robotics at a high level.

However, the current domestic trends are in the opposite direction, which is why the survey could be considered successful even if the overall average of the converted value of the responses were stagnant. This would show that the interest and desire for further occupations would not wane, and in my opinion, this could be an influential factor in the direction of students' further studies. Of course, the integration of robotics is also important to improve students' understanding of the material, but it can also be used to stimulate their attention and motivation.

However, it is important to note that natural science education should not rely exclusively on robotics. The lack of experimentation in the classical sense is also a problem. It is necessary to find the proportions and to consider when it is worthwhile to approach the material from a slightly different angle, rather than the usual experiments and videos. Therefore, there is no concrete suggestion as to the intervals at which these tools can be used during the lessons. In addition to the above, the structure of the curriculum may also influence the possibility of integration.

The feedback from the first part of the research shows that it was a success, and an encouraging sign that robotics does have a positive impact on students, just as much as non-conventional teaching and learning methods. These include cross-curricular interoperability and modern 21st century ways of working, which natural science lessons are a perfect synthesis subject, offer great opportunities for.

References

1. Sen, C., Ay, Z.S., Kiray, S.A.: Research Highlights in STEM Education (2018)
2. The 2018 International Computer and Information Literacy Study (ICILS): Main findings and implications for education policies in Europe
3. Patacsil, F.F., Tablatin, C.L.S.: Exploring the importance of soft and hard skills as perceived by IT internship students and industry: a gap analysis. J. Technol. Sci. Educ. **7**(3), 347–368 (2017)

4. Directorate general for internal policies policy department A: economic and scientific policy Encouraging STEM studies - Labour Market Situation and Comparison of Practices Targeted at Young People in Different Member State (2015)
5. Kennedy, T.J., Odell, M.R.: Engaging students in STEM education. Sci. Educ. Int. **25**(3), 246–258 (2014)
6. National Curriculum Framework for Natural Science Grade 5–6. https://www.oktatas.hu/pub_bin/dload/kozoktatas/kerettanterv/Termeszettudomany_5_6.docx
7. Eurostat - Graduates in tertiary education, in science, math., computing, engineering, manufacturing, construction, by sex - per 1000 of population aged 20–29. - 2012–2019
8. Gaál, B.: Possible ways of integrating robotics in natural science education. In: Szlávi, P., Zsakó, L. (eds.) InfoDidact2019. Webdidaktika Alapítvány, pp. 59–72. Zamárdi, Hungary (2019)
9. Aguilera, D., Ortiz-Revilla, J.: STEM vs. STEAM education and student creativity: a systematic literature review. Educ. Sci. **11**, 331 (2021). https://doi.org/10.3390/educsci11 07033
10. Academic research into the BBC micro:bit - micro:bit. https://microbit.org/research/
11. Meet the new BBC Micro:bit v2. https://microbit.org/new-microbit/
12. Keyestudio 37 in 1 Starter Kit for BBC micro:bit. https://wiki.keyestudio.com/KS0361(KS0 365)_keyestudio_37_in_1_Starter_Kit_for_BBC_micro:bit
13. Laal, M., Ghodsi, S.M.: Benefits of collaborative learning. Procedia Soc. Behav. Sci. **31**, 486–490 (2012)

Best Practice, Country, and Experience Reports

Clear the Ring for Computer Science: A Creative Introduction for Primary Schools

Marina Rottenhofer$^{(\boxtimes)}$, Lisa Kuka , and Barbara Sabitzer

Johannes Kepler University, 4040 Linz, Austria
{marina.rottenhofer,lisa.kuka,barbara.sabitzer}@jku.at
https://www.jku.at

Abstract. Nowadays, there is a high demand for qualified computer scientists and thus, it is important to start increasing computer literacy already at an early age. Although computer science and digital literacy are already becoming dominant topics and are increasingly anchored in the curricula, there is still a mismatch between what students and teachers understand the subjects to be and what they really are. In computer science, in particular, it is often believed that a computer is indispensable for teaching. The authors of this study want to change this picture by conveying basic computer science concepts, which play an essential role in many everyday situations. Thus, it is important to teach them from an early age. The challenge here is to find ways to present these fundamental ideas of computer science creatively so that young children, as well as laypersons, understand them. This paper introduces the COOL Computer Science Circus, which links selected core concepts with a circus show including interactive parts on side of a young audience. Alternatively, the show can be transformed into a workshop. The target group consists not only of primary school students aged 9–10, but also of primary school teachers. The authors describe their experiences and present first results of an ongoing survey which collects feedback to the COOL CS Circus as well as information about the teachers' interest and perception of computer science and digital education. Moreover, the paper gives an inspiration on how to present topics such as encryption, encoding, and algorithms without the need for a computer.

Keywords: Computer science · Creative computer science · Digital education · Primary education

1 Introduction

As computer scientists are needed in the workforce, it is essential to teach core concepts from an early age. However, these concepts can also be very handy in everyday situations. In everyday life, these concepts are often hidden or disguised, but at a closer look, it can be observed that so-called computational thinking (CT) skills are embedded in many areas. CT can be seen as a problem-solving thought process and thus the basis for computer science. Thus, it can be observed that digital skills, computational thinking, as well as computer science concepts find their way into school curricula worldwide and are taught at an early age. Having the opportunity to welcome a large group of primary school children at the COOL Lab of the Johannes Kepler University Linz each week for a half year, the authors had to come up with a creative idea to foster their interest in computer science. As the provided space for these events was the event hall "Circus of Knowledge" on campus, it inspired the idea to create a circus show. Thus, the challenge was not only to find a narrative form for computer science concepts that are suited for a young audience but also to provide interactive opportunities for a large group. This paper describes a way to present a narrative frame for the selected computer science concepts of encryption, encoding, binary numbers, and algorithms as well as to offer a form of interactivity for a young age group. Furthermore, the circus show was used to gain insights into the status of computer science in primary school resulting in the following research questions:

- How do primary school teachers and students perceive the presented circus show or Workshops?
- Are primary school teachers familiar with the presented computer science concepts?
- Do primary school teachers need further training to teach these concepts?

This is done via a questionnaire handed out to the teachers accompanying their students. In the following sections, a brief discussion of the development of the show is presented as well as a discussion on the preliminary results of the survey, as it is still going on.

2   Background

Since the focus is on developing a creative approach to introduce basic concepts of computer science to a young target group, this chapter illuminates the role of early computer science education in Austria, computer science core concepts, and their link to creative approaches in STEAM. Moreover, the COOL Lab as a meeting point for teaching and learning about computational thinking and computer science creatively is further explained.

2.1   Early Computer Science Education

In primary school, digital competencies are anchored in the curriculum. The focus is on media education and reflective use of the Internet, as well as a playful approach to technology and problem-solving [5]. Jeannette Wing argued that "computational thinking is a fundamental skill for everyone, not just for computer scientists. To reading, writing, and arithmetic, we should add computational thinking to every child's analytical ability" [12]. Thus, it can be said that it is necessary to teach computational thinking concepts from an early age on. Computational thinking and computer science education go hand in hand, since "Computational Thinking involves solving problems, designing systems, and understanding human behavior by drawing on the concepts fundamental to computer science" [12]. Also, studies confirm that computer science concepts can be taught successfully at an early age when the learning material and teaching are created age-appropriately [11].

2.2   Link Between Computer Science Core Concepts and Creative Approaches in the Field of STEAM

For the planning of the circus show, the authors analyzed several computer science core concepts and aimed to find links between these concepts and narrative situations as well as interactive tasks for a young audience. Four core ideas are encryption, encoding, algorithms, as well as binary numbers. Computers work with electrical signals, represented as 0 for power off and 1 for power on. This binary system is the very basis of a computer system. Encryption is a key-dependent conversion of data called "plaintext" into "ciphertext". The plaintext can be recovered from the ciphertext only by using a secret key. Encoding, on the other hand, is the process of translating particular data, for example, letters, numbers, punctuation marks, or symbols, into a special format so that it can be transmitted or stored more efficiently. Often a codebook is used for the encoding-decoding process. Whereas the focus of encryption lies in a secret and secure transmission of data, the main purpose of encoding is the protection of the integrity of data and maintaining its usability. The binary system is often seen as a code, for example. These concepts are often part of computer science curricula. An algorithm is often described as a step-by-step guide or recipe. Usually, its definition is based on five properties - executability, determinism, determinacy, finiteness, and termination. Each step must be executable. There is not only a finite number of steps but also always only one next step which is always uniquely determined. The algorithm itself must also end and produce a result that is always the same for the same input. Defining an algorithm in as many precise steps as necessary, but as few as possible is an important skill [9]. Algorithmic thinking is often described as a part of computational thinking and, thus, the basis for computer science curricula as well. When learning about these concepts at an early age, the focus does not lie on knowing the technical jargon, but rather on putting the knowledge into practice and fostering skills.

Studies have shown that science shows can increase the interest of the audience on scientific topics as well as educate them on the topics that e.g. include computer science [1]. The use of magic tricks is also an engaging approach to foster curiosity in mathematics and computer science alike [7]. Creative computer science activities can motivate young students and hopefully lead to the idea that they can become digital creators themselves, rather than to stick to passive consumer behavior [8].

Computer Science Unplugged (CS Unplugged) is an approach to conveying basic computer science concepts to students without the use of a computer, for example, by enacting algorithms by students themselves or using games and magic tricks. The activities created include detailed descriptions and use computational logic in an unplugged environment. It was first introduced by the University of Canterbury in New Zealand and has recently been adopted internationally by finding its way into school curricula worldwide. [2,3,13].

What previous approaches have in common is their interactivity. Constructivism is a well-known learning theory that proclaims that learning is not done via receiving and storing information from teachers, but rather that knowledge is constructed by students themselves. Thus, learning is an active process, rather than a passive one [4]. This also shows the main problem of the idea of packing computer concepts into a play. The solution lies in interactivity. Taking ideas from science shows, magic tricks, and hands-on material should motivate a rather passive theater audience to interact with the actors transforming them into active members of the cast of the play. For example, one of the Computer Science Unplugged activities transforms a parity-based error-correcting algorithm into a magic trick allowing the magician to identify a change in a card set that could not possibly be seen by them [6]. As magicians are often part of circus shows as well, this trick can be easily incorporated into a narrative for the circus.

2.3   JKU COOL Lab

In 2017, the JKU COOL Lab was developed at the STEM education department of the Johannes Kepler University Linz as a meeting point for not only teaching and learning but also research and practice. It is designed to foster digital literacy and computational thinking through creative approaches. On the one hand, it is open for students of all ages starting at kindergarten level offering workshops and clubs for (gifted) children, on the other hand, it provides practice opportunities for pre-service teachers as well as in-service training for experienced teachers. Interdisciplinarity is another aspect that is held high in the COOL Lab by developing, testing, and evaluating cross-curricula materials. Thus, using the COOL Lab resources and experience enhanced the process of creating the COOL Computer Science Circus. The predominant computer science expertise paired with pedagogical and didactic know-how was key to the successful implementation of the circus [10].

3   The COOL Computer Science Circus

The COOL computer science (CS) circus is a project developed by the COOL Lab of the Johannes Kepler University Linz, Austria. The COOL Lab is a teaching and learning lab for students of all ages as well as (prospective) teachers and focuses on computer science, computational thinking as well as digital education in general. The circus was developed for primary school children as well as teachers and aims to convey core concepts of computer science in a creative way and to arouse the interest of the participants. The premiere of the COOL

CS Circus took place at the beginning of March 2022 at the "Circus of Knowledge", the event hall of the Johannes Kepler University, and is currently being performed three times a week as part of the "Linz Aktion" (Linz campaign). This campaign offers 4th-grade primary school children the opportunity to get to know the diversity of Linz from different perspectives and shows them the most important sights and hotspots of the city in a trip of one or two days. The campaign takes place annually from February to the end of June, however, this year the start was postponed to March due to COVID-19 restrictions. Depending on the location at the JKU, the COOL CS circus is offered in two different variations, as circus show or workshop. The following subsections present the main demonstrations that are included in the 1-hour circus show and workshop and describe the core concepts that are conveyed to the students.

Fig. 1. Magic card trick

3.1   Let's Talk Binary

The circus show as well as the workshop start with the circus director who welcomes the children by saying "hello" in binary language, not knowing that they do not understand him. Fortunately, the magician Merlina and the bear are there in time to stop the circus director calling out ones and zeros and explain to the children why he is talking so funny. In easy words, the bear talks about the role of binary language in computer science.

3.2   Toast, Chocolate Cream and Magic Tricks

In the next part of CS Circus, the magician Merlina gives the participants an insight into detecting errors and algorithmic thinking by showing them some magic tricks. In the show, Merlina asks for a volunteer to come on stage and support him while he performs a card trick. With this magic trick, Merlina can tell exactly which card the volunteer child turned over, even though she turned around and saw nothing. This demonstrates one of the techniques computers are using to detect data errors automatically (Fig. 1). In the CS Circus workshop, on the other hand, not only the card trick is presented, but also the chocolate cream challenge to teach children the concept of algorithms. In this challenge, the participants have the task to instruct Merlina to prepare a toast with chocolate cream on it. The task sounds easy at the beginning, however, the children soon recognize that many things can go wrong when making a simple sandwich if the instructions aren't clearly formulated.

3.3   Save the Bear

Since the bear blew up Merlina's magic trick by explaining the logic behind it, Merlina cast a spell on the bear out of anger. Now the task of the children is to free the bear by cracking the code of the lock and to give the clown clear directions to get to the bear. The children manage to do this by solving a computer science quiz and decoding the correct answers with the Caesar cipher. With this approach, participants are introduced to the concept of encryption and apply again algorithmic thinking by giving the clown clear instructions. In the workshop, the participants also work with the Caesar cipher to solve various tasks (Fig. 2).

Fig. 2. Save the bear

3.4   Tame the Bees and Dance

The children perfectly manage to save the bear and all members of the circus are very relieved that the story has a happy ending thanks to the children. To celebrate, the circus members demonstrate "dance programming" to the children and teachers. With the help of various symbols, a dance is rehearsed, and they then all dance together. During the workshop, the ringmaster acts as a bee tamer and asks the children to solve various tasks with the Bee-Bots, which are small robots that help to train estimation, problem-solving, and sequencing (Fig. 3).

Fig. 3. Dance programming

4   Methods and Results

To receive feedback on the COOL CS circus and to gain insights into primary school teachers' experiences and attitudes towards computer science and digital education, a questionnaire was developed and is currently being distributed to all the teachers who participate in the show or workshop as part of the "Aktion Linz". The questions of the survey used a five-point Likert scale including (1) not true, (2) rather not true, (3) partly true, (4) rather true, and (5) true. The survey also included open-ended questions, to receive suggestions for improvement of the COOL circus as well as to find out more about the role of computer science and digital education in school. Demographic variables included gender and years of service and the survey was completed anonymously including a unique identifier. The questionnaire was issued in German and translated for this paper. For the statistical analysis, the software IBM SPSS Statistics 23 was used.

This paper presents the preliminary answers to the survey related to the COOL circus starting from the first performance at the beginning of March until mid-May 2022. Until that time, the questionnaire was administered to a total of 109 teachers who visited the circus with a total of 1030 children. The 27.5% of the teachers who responded to the questionnaire consisted of 1 male and 29 females with an average of 14.8 years of service and a standard derivation of 10.36. Of those teachers, 9 attended the circus performed as a show at the JKU's event hall "Circus of Knowledge" with their classes and 21 the workshop.

Table 1. Descriptive statistics.

Item	N	Mean	Std. derivation
I liked the offer "Clear the Ring for IT"	30	4.47	.819
I have the feeling that the offer was well received by my students	28	4.64	.559
The contents of the offer were understandable for the students	29	4.38	.862
I already knew the IT concepts and content of the offer (e.g. algorithm, coding, encryption...)	28	3.14	1.177
The content has already been covered in class	29	1.62	.942
I can imagine implementing this content WITHOUT training in my own classes	28	2.21	1.134
I can imagine implementing this content WITH training in my own courses	29	3.52	1.184
I think it makes sense to convey computer science concepts creatively, e.g. with movement and dance	29	3.93	.998
The image of computer science has changed for me after the show/workshop	27	3.30	1.171

The responses to the nine Likert items related to the COOL CS Circus are visible in Table 1. Overall, the offer was well received by the teachers with a mean score of 4.47 and a standard derivation of 0.819. Furthermore, 28 teachers indicated that they have the feeling that the offer was also well received by their students (mean= 4.64, SD= .559) and that the content was understandable for them (mean = 4.38, SD= .862). The following items referred to the teachers' prior knowledge and the implementation of CS in class. When asking whether the teachers already knew the IT concepts and the content of the offer (e.g. algorithm, coding, encryption...), 13.3% said "not true", 6.7% "rather not true", 36.7% "partly true", 26.7% "rather true" and 10% "true". Even though this indicates that more than 70% is at least partly familiar with the concepts, the content is hardly implemented in class (56% "not true", 26,7% "rather not true", 10% "partly true", 3.3% "true"). The responses to the questions "I can imagine implementing this content WITH/WITHOUT training in my own classes" are shown in Fig. 4. Without training, none of the teachers can fully imagine implementing IT concepts without training in class, 11 of them indicated that they would not implement them at all without support. Comparing these answers with "I can imagine implementing this content WITH training in my class" it is visible that many teachers are open to incorporating computer science concepts into their own lessons (7 indicating "true", 9 "rather true", 6 "partly true", 6 "rather not true" and 1 "not true").

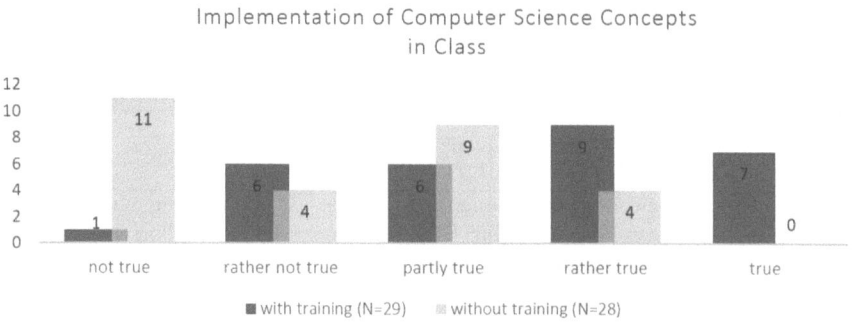

Fig. 4. I can imagine implementing this content with/without training in my own class.

The Likert items were followed by one open-ended question to receive feedback for improvement of the COOL CS Circus. The response rate was 36,7% including plenty of praise for this offer but also constructive criticism. Two of them were related to the space, noise level, and the obligation to wear masks. For the workshop, we are allocated different rooms on campus, which are sometimes not ideal for implementation and thus lead to increased noise levels. Also, due to COVID-19 regulations, wearing an FFP2 mask was still mandatory, so, it was sometimes difficult to understand the actors. In terms of content one teacher had wanted a preparation unit for her own lessons before the performance and similarly, another teacher noted that the term computer science should be brought closer to the children beforehand. One teacher suggested incorporating additional material from other institutions and another teacher would wish that this program could also be offered directly in schools.

5   Conclusion

The COOL CS Circus has demonstrated that it is possible to convey computer science concepts in narrative form with an interactive circus show as well as in the form of a workshop. Preliminary results of a survey dedicated to primary school teachers indicate that computer science concepts are not yet very well represented in primary school, but that teachers are open to this topic. Many participants in the survey can even imagine incorporating the content into their lessons themselves, others would like more external offers and materials on this topic as well as support. Furthermore, teachers have a positive attitude towards the use of creative elements to teach computer science.

References

1. Bell, T.: A low-cost high-impact computer science show for family audiences. In: Proceedings 23rd Australasian Computer Science Conference. ACSC 2000 (Cat. No. PR00518), pp. 10–16. IEEE (2000)
2. Bell, T., Alexander, J., Freeman, I., Grimley, M.: Computer science without computers: new outreach methods from old tricks. In: Proceedings of the 21st Annual Conference of the National Advisory Committee on Computing Qualifications, pp. 127–133 (2008)
3. Bell, T., Alexander, J., Freeman, I., Grimley, M.: Computer science unplugged: school students doing real computing without computers. N. Z. J. Appl. Comput. Inf. Technol. **13**(1), 20–29 (2009)
4. Ben-Ari, M.: Constructivism in computer science education. Sigcse Bull. **30**(1), 257–261 (1998). https://doi.org/10.1145/274790.274308
5. BMBWF: Digitale Grundbildung - Digitale Grundbildung in der Primarstufe - Bundesministerium für Bildungswissenschaft und Forschung (2022). www.bmbwf.gv.at/Themen/schule/zrp/dibi/dgb.html#:~:text=DigitaleGrundbild unginderPrimarstufe,ZugangzuTechnikundProbleml{\"o}sung
6. Computer Science Unplugged: Parity magic. https://www.csunplugged.org/en/topics/error-detection-and-correction/unit-plan/parity-magic/
7. Curzon, P., McOwan, P.W.: Engaging with computer science through magic shows. In: Proceedings of the 13th Annual Conference on Innovation and Technology in Computer Science Education, pp. 179–183 (2008)
8. Giannakos, M.N., Jaccheri, L., Proto, R.: Teaching computer science to young children through creativity: lessons learned from the case of norway. In: CSERC, pp. 103–111 (2013)
9. Hare, K.: Computer Science Principles: The Foundational Concepts of Computer Science, 4th edn. Yellow Dart Publishing, Atlanta (2022)
10. Sabitzer, B., Demarle-Meusel, H., Painer, C.: A Cool Lab for Teacher Education. Teacher Education for the 21st Century, p. 319 (2019)
11. Schwill, A.: Ab wann kann man mit Kindern Informatik machen? Eine Studie über informatische Fähigkeiten von Kindern. In: Keil-Slawik, R., Magenheim, J. (eds.) Informatikunterricht und Medienbildung, INFOS 2001, 9. GI-Fachtagung Informatik und Schule, pp. 13–30. Gesellschaft für Informatik e. V., Bonn (2001)
12. Wing, J.M.: Computational thinking. Commun. ACM **49**(3), 33–35 (2006). https://doi.org/10.1145/1118178.1118215
13. Wohl, B., Porter, B., Clinch, S.: Teaching computer science to 5–7 year-olds: an initial study with scratch, cubelets and unplugged computing. In: Proceedings of the Workshop in Primary and Secondary Computing Education, pp. 55–60 (2015)

Bebras Tasks Based on Assembling Programming Code

Jiří Vaníček$^{(\boxtimes)}$ ⓘ, Václav Šimandl ⓘ, and Václav Dobiáš ⓘ

University of South Bohemia, České Budějovice, Czech Republic
{vanicek,simandl,dobias}@pf.jcu.cz

Abstract. The paper examines the creation and evaluation of so-called situational informatics tasks based on assembling a program from blocks. Blockly technology has enabled us to develop an environment where templates, called "worlds", can be created. In these worlds, pupils program a certain sprite to solve a problem emerging in a described situation. We created two such templates – the world of Karel the robot and the world of Film animation, differing both in behavior of sprites and set of commands. Each template was supplied with its own set of tasks, differing in topic, subject matter and graphics. As they go through each task, pupils repeatedly run the assembled program, being provided by the system with feedback. That comprises a visual check of how the programmed sprite behaves as well as system-generated notifications reporting whether all the requirements for completing a task have been met. The tasks that were compiled for this purpose were included in the Bebras Challenge. In our paper, we describe each of the templates and look at their didactic background as well as examining findings from the practical use of these tasks in the Challenge and their inclusion in the informatics curriculum. Results show that tasks created for the world of Karel the robot used in the Bebras Challenge are no more difficult than other algorithmic tasks. Moreover, informatics teachers are impressed with these tasks and they find it of upmost importance that the curriculum includes such tasks in order to advance pupils' informatics skills.

Keywords: Computational thinking · Algorithmization · Block programming · Primary school · Secondary school · Bebras Challenge

1 Introduction

Programming is generally perceived as a matter of specialized professional training. However, Gander claims it is an essential part of general public education for the 21st century [1]. Not only providing us with the opportunity to discover the world from another perspective and understand how a computer works, programming can also be perceived as a microworld that can develop an individual's mental abilities. The direction of teaching and subsequent choice of appropriate educational content, environment, motivation and teaching methods all derive from its basic definition. From the perspective described above, programming is presented in our article as a training ground for

A. Bollin and G. Futschek (Eds.): ISSEP 2022, LNCS 13488, pp. 113–124, 2022.
https://doi.org/10.1007/978-3-031-15851-3_10

developing an individual's abilities and competences. The same approach is found in strategic documents like the "Shut down or restart?" study in the United Kingdom [2], the worldwide ACM computing curriculum [3], CSTA K-12 computer science standards in the United States [4], our close neighbors' Štátny vzdelávací program in Slovakia [5] and Podstawa programowa z informatyki in Poland [6].

If we consider education's general aim as being to primarily develop personality, in the field of informatics it is the development of computational thinking [7] as the ability to find a solution to a problem in a form which could be automatically carried out by an information-processing agent. Algorithmization, i.e. identifying a method or process to achieve a goal and formulating it in a way so that such an agent could read and perform it, is a fundamental part of computational thinking. This term became the mainstay for defining school curriculum content in the above-mentioned countries as well as the Czech Republic in its government Strategy of Digital Education [8] and in its proposal for new General Curriculum programs [9].

According to Cuny, Snyder and Wing, the idea of computational thinking for everyone involves abilities such as understanding what aspects of a problem are amenable to computation, evaluating the match between computational tools and techniques and a problem, using or adapting a computational tool for a new use and identifying opportunities to use computation in a new way [10]. In order to be able to develop these abilities through programming, we must try to find suitable approaches, situations, tasks and also environments which will highlight and emphasize these goals. Wittmann defines such an environment as a set of interconnected situations providing problems which enable a pupil to identify important thoughts [11].

Xia defines teaching of programming as supporting students to understand the concepts of programming via hands-on experiences and learning as the activity of obtaining useful programming knowledge and skills by studying [12]. Many approaches to teaching programming favor student activity, active learning, learning by doing, and the construction of knowledge as a result of active creative work. All this with respect to the fact that knowledge and knowing are not transmittable. According to Piaget, knowledge is actively constructed by the learner in interaction with the world, so, as Ackermann [13] suggests, it is worth providing opportunities for children to engage in hands-on explorations that fuel the constructive process. Ackermann quoted Piaget's theory that "children interpret what they hear in the light of their own knowledge and experience", and his belief that "knowledge is formed and transformed within specific contexts, shaped and expressed through different media" [14]. How one constructs knowledge is a function of the prior experiences, mental structures, and beliefs that one uses to interpret objects and events [15].

Two basic types of programming tasks can be found in coursebooks and manuals for the teaching of programming:

- "Études" lasting several minutes, always focused on a specific skill or programming concept, their aim being particular knowledge acquisition;
- Bigger "projects", often in the form of creating stories or games which are more complex, the outcome being a product.

Études allow better detection of a learner's error and appropriate sorting of such tasks facilitates the creation of mental models of a learner. Projects require a combination of more skills at the same time, including planning and creativity; the created longer programming codes require more knowledge from a learner but this is counterbalanced by higher motivation, as a learner works towards a final product. For example, études were used in the textbook [16], projects were used in the textbook [17].

If a learner solves problems by creating software, specifically by assembling a program, a teacher can get feedback by analyzing the written program to determine how a learner has understood the situation which the problem he/she is solving occurs in; how well he/she understands the concepts he/she uses; what level he/she has reached in terms of elements of computational thinking like algorithmization, decomposition and generalization; or the approaches he/she uses to solve problems [18].

The Bebras Challenge has contributed to the development of computational thinking for a number of years [19], being held in more than 60 countries worldwide [20]. Via an online test, the tasks put learners in a situation where they have to determine the corresponding informatics concept and select an answer by applying their computational thinking. The situational tasks used in the Bebras Challenge – in the Czech version called Bobřík informatiky [21] – are similar to études in their structure and focus on a particular informatics concept. The contest consists of an online test so there are multiple-choice, click on object or drag object tasks. Bebras produces a number of new informatics tasks, contributing to innovations in the school curriculum, some of the tasks being incorporated into new Czech informatics coursebooks [22].

1.1 Motivation and Aim

Situational "Bebras" tasks should develop various aspects of computational thinking including algorithmization. Typical algorithmic tasks used in the contest include identifying start and end state after applying a particular algorithm, comparing several algorithms with a task assignment, considering rules for carrying out a computation, identifying an error in an algorithm or its optimalization. The contest did not include tasks to be answered by assembling a program, which restricted the variety of informatics tasks in the contest.

We tried to find a solution that would enhance the existing contest to include tasks where, just like in the programming environments used in schools, contestants could assemble a program from blocks. This would involve using the widespread concept of block programming, known from programming environments like Scratch, Blockly or MakeCode, which learners are familiar with from informatics or robotics coursebooks. Such a solution will bring innovation into the Bebras Challenge which will benefit from this newly developed type of tasks.

2 Methods

Solution methodology proceeds from design-based research as per Trna [23]. To develop the software module, we first analyzed familiar open source block programming environments (e.g. Scratch, Blockly, MakeCode) to determine whether they could be used to create the software module, primarily in terms of pedagogy and implementation. We then analyzed familiar "worlds" which the programming tasks would be created in (e.g. Karel the Robot, turtle graphics, Baltie the magician) in terms of:

- their ability to cover the curriculum range (minimum of pre-entry knowledge, maximum of educational aims),
- their suitability for the creation of a set of tasks that progress in small steps with regards to acquired knowledge
- their suitability for the creation of particular "contest" tasks, considering the specific nature of each one.

2.1 Design

We designed and developed a software environment where interactive situational programming tasks can be created. A learner solves a problem in it by assembling a program from blocks and is given the possibility of testing and debugging his/her program. We implemented a modified module of Blockly [24] into our environment.

The advantage of block programming is that it prevents syntax errors. In our implementation, it also has a limited set of programming commands, which encourages learners to think rather than searching for a tool that could conveniently solve the problem for them.

As they go through each task, pupils repeatedly run the assembled program, being provided with feedback by the system. That comprises a visual check of how the programmed sprite behaves as well as system-generated notifications as to whether all the requirements for completing a task have been met. The environment enables the creation of sets of tasks that follow on as the learner progresses, similar to Hour of Code activities [25], resulting in tasks of increasing difficulty with the progressive employment of more complex concepts and situations.

Each time a learner asks for a program to be run by clicking on Run button, the system simultaneously saves the learner's program along with information as to whether his/her solution has met all assigned requirements and the number of attempts the learner required to create the right program.

To accompany this software environment, we developed two templates of programming tasks, so-called "worlds", each having program-controlled sprites that behave differently and each having different sets of basic commands. Each template was supplied with its own set of tasks, differing in topic, subject matter and graphics.

1st World: Controlling the Robot.

The first template simulated "the world of Karel the robot" [26], a sprite that walks around a system of squares picking up and putting down objects. The basic set of commands can move the programmed sprite around the game board one square at a time, make a quarter turn in both directions, detect objects on the square where the sprite is located and remove such an object from a square or place an object on a free square. It can also detect an obstacle on an adjacent square in the direction the sprite is facing. The basic language commands were supplemented with a Repeat structure, constituting a loop with a fixed number of repeats (see Fig. 1).

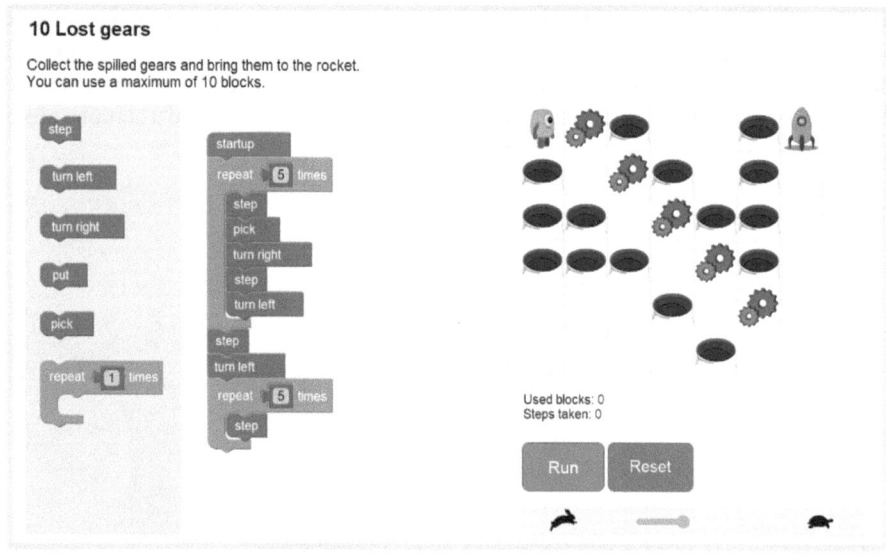

Fig. 1. The world of Karel the robot and the task in which learners are assigned to collect gears. The program created by a learner to complete the task is in the middle.

We chose "the world of Karel the robot" as it is simple enough for a learner to understand and manage the basics of the language so that he/she can quickly move on to more complicated tasks. The advantages of this "world" include the possibility to create real situations, the visual clarity of the sprite's status, facing one of the four main directions so not requiring a turning angle parameter as well as the limited number of basic language commands (step forward, turn, pick, put). Another advantage is the absence of more complicated terms like object, coordinates, procedure or variable. This reduces the amount of time needed to get acquainted with the environment, meaning that learners can soon progress onto and concentrate more intensively on the algorithmic core of the solved problems.

Typical tasks in this world were to go to a particular place, avoid an obstacle, pick up equally distributed objects or find a way composed of multiple parts. In this world, we have created a set of programming tasks which gradually increase in difficulty. The loop programming concept was used to repeat one block, assemble a program with blocks preceding and/or following a loop; with several blocks in the loop body; with several loops following each other in a program and find a way to complete a task using the shortest possible programming code. From the 4th task on, the number of blocks that could be used in a program was limited, forcing learners to shorten code and use the loop.

2nd World: Animation

The other developed template was the so-called "World of Film", in which a sprite is programmed to change its position and size over time. The sprite has four parameters: position X, position Y, size and rotation (as opposed to one basic direction). The programming language has one basic command Sprite, which draws a sprite on the game

board in the place given by the parameters of X, Y positions, its size given by a parameter and rotated by a given number of degrees. This command was supplemented with a
block for creating mathematical expressions with basic arithmetic operations and the If
structure controlling the time condition (e.g. whether time has exceeded a certain value).

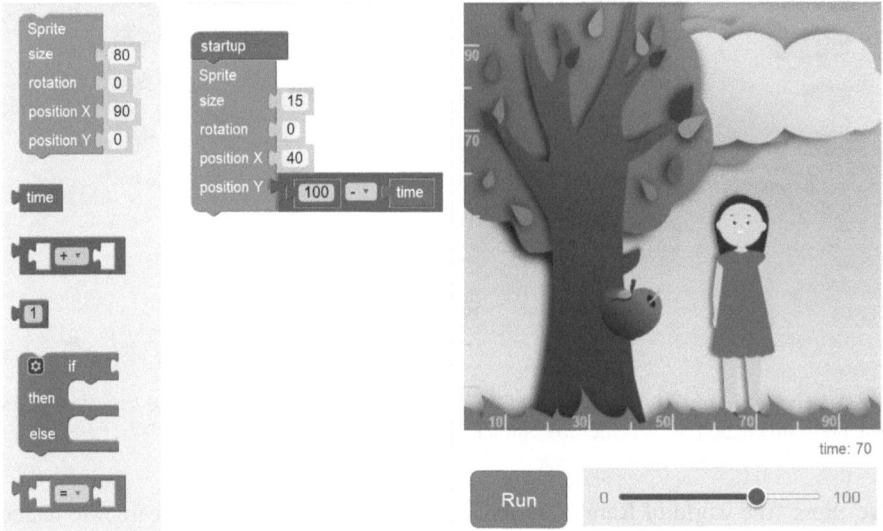

Fig. 2. The World of Film. Learners are assigned to animate an apple falling from a tree (the
situation in the picture having a time value of 70). The correct solution to the task can be seen in
the middle, the time variable having been used in the expression.

Animation is carried out in such a way that when the program is run, the time variable
continuously changes its value from 0 to 100 and the Sprite command is performed for
each of these values. The time variable can be used as a parameter in a command. If the
value of the X position parameter is set equal to the time variable, the X position will
continuously change from 0 to 100 and the sprite will move uniformly from left to right
over the whole game board.

  The learner is assigned the programming task by running animation of the programmed sprite's shadow. The learner's task is to create a program (i.e. assemble parameters of the Sprite block) to make his/her sprite behave in the same way as its shadow,
i.e. both objects should overlap each other throughout the animation (see Fig. 2). The
learner has the possibility of running the program repeatedly, animation having the time
value of 0 to 100. He/she can also use a scroll bar to manually set any value of the time
variable and analyze the situation at a given time (see Fig. 3).

  Whereas the didactical aim of the world of Karel the robot is to get fluent with loops,
the world of Film aims to understand procedures with parameters. The world of Film is
based on the parameter concept, working with the variable and primarily with expressions. The method of programming in this world is close to functional programming.
Consisting of more complicated concepts than the world of Karel the robot, it is more
suitable for learners at high school or in their final years of lower secondary school.

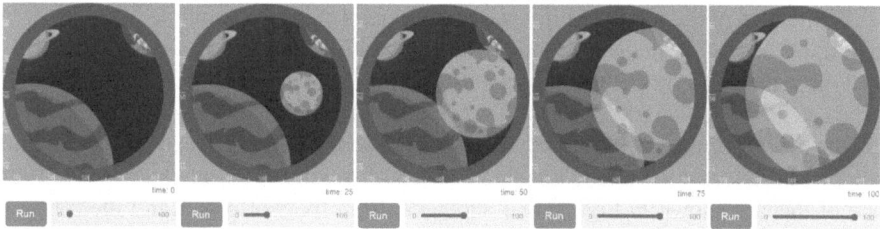

Fig. 3. Phased animation of a task requiring a spaceship to land on a planet in time values of 0, 25, 50, 75, 100, showing the shadow of the planet getting closer and bigger over time.

In this world, learners were typically assigned to move a sprite horizontally or vertically in uniform motion (motion from right to left is more complicated than motion from left to right due to position becoming smaller as against time), to move it around more slowly and more quickly, to combine motion in those directions, to make the sprite grow or shrink over time and to combine growth with the motion of the sprite. In this world, we created a set of programming tasks which gradually increase in difficulty. More simple tasks include placing a sprite in a specific position in the coordinate system or increasing one of its coordinates in relation to time. More complicated tasks include the use of expressions to decrease one of the sprite's coordinates or its size as time increases, the combination of several parameters dependent on time (e.g. motion on the diagonal or simultaneous motion and shrinking of a sprite). The most difficult tasks combined several motions over time (e.g. motion there and back during one animation), applying decision-making.

2.2 Evaluation

The created environment for assembling programs from blocks was implemented as a module in the Czech edition of Bebras Challenge. We used the created tasks in two ways.

As programming is not a compulsory part of Czech Informatics curricula, we supposed that many pupils have no programming skills. Thus, we created a special set of tasks called Blocks which contained tasks from the world of Karel the robot. We offered this set of tasks to schools as preparation for the national round of the contest. During September and October 2021, this test of 11 questions was taken by 45 000 learners at lower secondary and high schools. Having examined findings drawn from feedback from schools and from a consulting expert's review, we made improvements to the environment and tasks. Problems with graphics not working properly in some tasks in some browsers were most common. There were also reports of difficulties in transition between task 4 and 6, caused by a very large cognitive step. We solved this problem by inserting another task 5 and adding explaining elements to the task questions.

The national round of the Bebras Challenge was another iteration for verification. Each of the 109 442 contestants worked on 3 completely new tasks from the world of Karel the robot (the total number of tasks being 12). Tasks in older age categories were based on more complex algorithmic situations. This iteration allowed us to determine to what extent these new tasks are more difficult than other algorithmic tasks and to what extent they are more difficult than the average task (see Results for further details).

Verification of the world of Film was also carried out in two iterations, despite fewer contestants having participated. During January and February 2022, a set of 12 tasks of this type served as a practice set for contestants that had qualified for the central round in the category for the oldest pupils. 546 learners worked on this set of tasks.

The second iteration was carried out in the central round itself, which 358 contestants took part in. The test was made up of 15 tasks, 3 of which were from the world of Film (one of them is shown in Fig. 4) and another 3 from the world of Karel the robot. It means that 40% of the tasks involved programming by assembling programming code from blocks. Verification showed that these tasks can be used at such a high level as the central round of a nationwide contest.

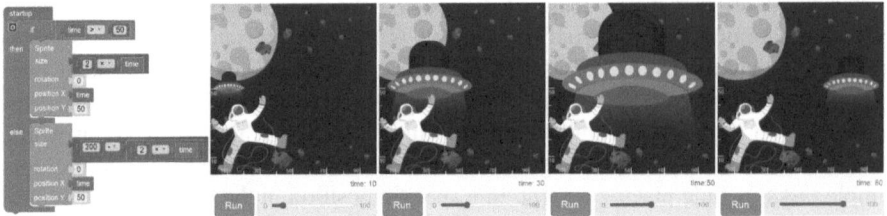

Fig. 4. A complex task where the sprite first moves closer and then moves away, taken from the central round of the contest, using program branching depending on the time parameter. The correct solution being on the left, its phased animation is on the right in time values of 10, 30, 50 and 80 (at time 0 and 100, the flying saucer measured zero).

Following verification, we had to improve the method of the learner's data evaluation while solving a task. There were deviations in computations when using parameters in combination with multiplication and division (e.g. multiplication by 0.001 and division by 1000 did not give the same result), which, in rare cases, led to incorrect evaluations of learners' solutions.

It also emerged that rotating a sprite visually by angular degrees is not optimal as it had rotated the sprite by 100° by the end of the animation. Learners can easily recognize a 90° rotation but for the sprite to rotate by 90° during a time interval of 100, it would have to be rotated in grades rather than in angular degrees, i.e. in units that learners are not acquainted with in schools. For that reason, we finally decided not to use the rotation parameter in tasks.

3   Results

In order to compare the difficulty of programming tasks involving assembling blocks as against other algorithmic task, we worked with proportion of contestants that had been able to solve a task, which is the factor for describing task difficulty [27]. While devising tests for the Challenge, we made efforts to create programming tasks which difficulty coincided with the overall difficulty level of algorithmic tasks. A verification process was used to ascertain whether we had managed to do so.

First, we used the Anderson-Darling test to ascertain whether success rates for programming tasks and other algorithmic tasks are normally distributed. The hypothesis for testing normality of data was rejected at a significance level of 0.05. We then tested the equality of variances of both samples. At the level of 0.05 the null hypothesis was not rejected. Therefore, it can be claimed that variances of both samples are equal.

We subsequently used the two-sided non-parametric Wilcoxon test to ascertain whether the means of both samples are equal. Since the null hypothesis was rejected at a significance level of 0.05, we used the one-sided non-parametric Wilcoxon test. This enabled us to verify whether the mean of the success rate for the programming tasks was equal to or lower than for the other algorithmic tasks. As this hypothesis was rejected at a significance level of 0.05, it can be claimed that the success rate for programming tasks involving assembling blocks is significantly higher than for other algorithmic tasks.

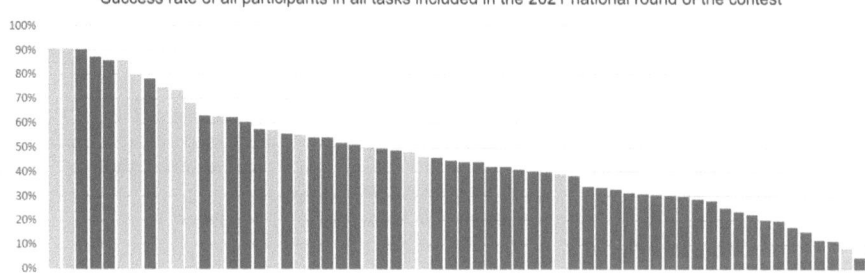

Fig. 5. A comparison of the difficulty level of all tasks included in the 2021 national round of the contest; tasks that involve assembling programming code are marked in a light color.

When comparing programming tasks with all contest tasks in this year's national round, we discovered that, apart from two exceptions, all of these tasks from all age categories were placed in the top half according to the proportion of correct answers (see Fig. 5). In this figure, programming tasks involving assembling blocks are marked in a light color. To confirm a statistically significant difference in the difficulty of programming tasks as against other task, a verification process was used.

First, we used the Anderson-Darling test to ascertain whether success rates for programming tasks and other tasks are normally distributed. The null hypothesis for testing normality of data was not rejected at a significance level of 0.05. We then tested the equality of variances of both samples. At a significance level of 0.05 the null hypothesis was not rejected. Therefore, it can be claimed that variances of both samples are equal. We subsequently used the two-sided Student's t-test to ascertain whether the means of both samples are equal. Since the null hypothesis was rejected at a significance level of 0.05, we used the one-sided Student's t-test. This enabled us to verify whether the mean of the success rate for the programming tasks was equal to or lower than for the other tasks. As this null hypothesis was rejected at a significance level of 0.05, it can be claimed that the success rate for programming tasks is significantly higher than for other tasks.

It means that programming tasks involving assembling blocks can be declared as being demonstrably easier for learners. This may be due to the fact that the tasks provided feedback and learners could have several attempts to iterate their answers, unlike in regular Bebras tasks, where they click on objects or answer multiple choice questions without receiving any feedback. The attractiveness of this new type of tasks is another factor that has to be taken into account.

In November 2021 we asked 939 school coordinators who were responsible for organizing the contest in their school to fill in a questionnaire. 199 replies were received, representing a 20% rate of return. Therefore, it can be regarded as a statistically representative sample and the views expressed can be taken into consideration.

One of the questions referred to how teachers rated the new type of Bebras tasks, based on assembling a program from blocks. Using the Czech school grading system, almost three quarters of them rated this type of tasks "excellent", one sixth "very good" and the remaining tenth "good", "satisfactory" or "unsatisfactory".

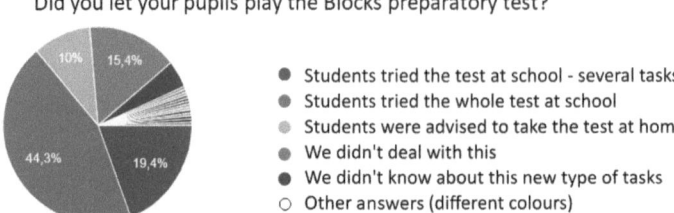

Did you let your pupils play the Blocks preparatory test?

- Students tried the test at school - several tasks
- Students tried the whole test at school
- Students were advised to take the test at home
- We didn't deal with this
- We didn't know about this new type of tasks
- Other answers (different colours)

Fig. 6. Results of the poll question asking teachers whether and how they gave their learners the opportunity to play the Blocks preparatory test in the world of Karel the robot.

Another question inquired into the extent teachers used their lessons to give learners the opportunity to work on the type of tasks used in the world of Karel the robot. Results show that almost two thirds of schools involved in the poll had given pupils the opportunity to prepare for the contest during lessons (for details see Fig. 6). Schools can therefore be regarded as perceiving this type of tasks to be an appropriate innovation to the curriculum for advancing learners' informatics skills.

4   Conclusion

We introduced a new type of informatics tasks into the Czech edition of the Bebras Challenge, not having previously been used in that informatics contest. The tasks involve the use of blocks to assemble programming code. Considering the potential of the block environment, such tasks are of value both from a motivational and a pedagogical respect. Apart from their significance in the Bebras Challenge, they will also be of vital importance in the teaching of computing in primary and secondary schools. The developed tool can be used to create and test sets of tasks focusing on one concept or one programming skill, with the possibility of later implementing them into the school curriculum. In the future the developed module can be enhanced by adding more "worlds" such as turtle graphics or the world of Baltie the magician.

Being able to continuously save created programs, the environment allows monitoring of the way a learner deals with a programming task or the way he/she progresses in a set of tasks. That will enable future research to examine the programming mistakes a beginner might make, what kind of instructions might contribute to or eliminate their occurrence or to reveal misconceptions that might prevent beginners from solving programming tasks in a block environment. This could be useful for future compilation of the programming curriculum for beginners, where appropriate tasks could be chosen either to prevent typical mistakes or, contrarily, to lead a learner into making them, allowing the potential of failure to be used to develop a learner's understanding.

Acknowledgement. The research was supported by the project TAČR TL03000222 "Development of computational thinking by situational algorithmic problems".

References

1. Gander, W.: informatics and general education. In: Gülbahar, Y., Karataş, E. (eds.) ISSEP 2014. LNCS, vol. 8730, pp. 1–7. Springer, Cham (2014). https://doi.org/10.1007/978-3-319-09958-3_1
2. The royal society: shut down or restart? The Way Forward for Computing in UK Schools. The Royal Society, London (2012). https://royalsociety.org/~/media/royal_society_content/education/policy/computing-in-schools/2012-01-12-computing-in-schools.pdf
3. K-12 Computer Science Framework Steering Committee: K-12 Computer Science Framework. ACM, New York, NY (2016). https://dl.acm.org/doi/book/10.1145/3079760
4. CSTA: K-12 Computer Science Standards (2011)
5. Blaho, A.: Informatika v štátnom vzdelávacom programe (Informatics in a state educational programme). In: Kalaš, I. (ed.) DidInfo 2012, pp. 7–14. Matej Bel University, Banská Bystrica (2012). http://www.didinfo.net/images/DidInfo/files/didinfo_2012.pdf
6. Sysło, M.M., Kwiatkowska, A.B.: Introducing a new computer science curriculum for all school levels in Poland. In: Brodnik, A., Vahrenhold, J. (eds.) ISSEP 2015. LNCS, vol. 9378, pp. 141–154. Springer, Cham (2015). https://doi.org/10.1007/978-3-319-25396-1_13
7. Wing, J.M.: Computational thinking. Commun. ACM **49**(3), 33–35 (2006). https://doi.org/10.1145/1118178.1118215
8. Ministry of education, youth and sports of the Czech Republic: Strategie digitálního vzdělávání (Strategy of digital education). Ministry of education, youth and sports of the Czech Republic, Praha (2014). https://www.msmt.cz/uploads/DigiStrategie.pdf
9. Ministry of education, youth and sports of the Czech Republic: Rámcový vzdělávací program pro základní vzdělávání (Frame educational programme for basic education – basic version). Ministry of Education, Youth and Sports of the Czech Republic, Praha (2021). https://www.edu.cz/wp-content/uploads/2021/07/RVP-ZV-2021.pdf
10. Wing, J.M.: Computational thinking: what and why? Carnegie Mellon University, Pittsburgh (2010). https://www.cs.cmu.edu/~CompThink/resources/TheLinkWing.pdf
11. Wittmann, E.H.: Developing mathematics education in a systemic process. Educ. Stud. Math. **48**(1), 1–20 (2001). https://www.jstor.org/stable/3483113
12. Xia, B.S.: A pedagogical review of programming education research: what have we learned. Int. J. Online Pedagog. Course Des. **7**(1), 33–42 (2017). https://doi.org/10.4018/IJOPCD.2017010103

13. Ackermann, E.: Constructivism(s): shared roots, crossed paths, multiple legacies. In: Clayson, J.E., Kalaš I. (eds.) Constructionism 2010: Constructionist Approaches to Creative Learning, Thinking and Education: Lessons for the 21st Century: Proceedings for Constructionism 2010. Comenius University, Bratislava (2010)

14. Ackermann, E.: Piaget's constructivism, Papert's constructionism: what's the difference? (2001). http://learning.media.mit.edu/content/publications/EA.Piaget%20_%20Papert.pdf

15. Jonassen, D.H.: Objectivism versus constructivism: do we need a new philosophical paradigm? Educ. Tech. Res. Dev. **39**, 5–14 (1991). https://doi.org/10.1007/BF02296434

16. Kalaš, I.: UCL Scratchmaths curriculum. University College London, London (2017). http://www.ucl.ac.uk/ioe/research/projects/scratchmaths/curriculum-materials

17. The LEAD Project: Easy LEAD: Super Scratch programming adventure! No Starch Press, San Francisco (2012)

18. Chao, P.-Y.: Exploring students' computational practice, design and performance of problem-solving through a visual programming environment. Comput. Educ. **95**, 202–215 (2016). https://doi.org/10.1016/j.compedu.2016.01.010

19. Dagienė, V.: The bebras contest on informatics and computer literacy – students drive to science education. In: Joint Open and Working IFIP Conference, ICT and Learning for the Net Generation, pp. 214–223. Kuala Lumpur (2008). https://www.bebras.org/sites/default/files/documents/publications/DagieneV-2008.pdf

20. Bebras Challenge. https://www.bebras.org/

21. Bobřík informatiky (Beaver of Informatics). https://www.ibobr.cz/english-uk

22. Berki, J., Drábková, J.: Základy informatiky pro 1. stupeň ZŠ (Basic of informatics for primary school). Textbook. Technical University of Liberec, Liberec (2020). https://imysleni.cz/ucebnice/zaklady-informatiky-pro-1-stupen-zs

23. Trna, J.: Konstrukční výzkum (design-based research) v přírodovědných didaktikách. Scientia in educatione. **2**(1), 3–14 (2011). https://ojs.cuni.cz/scied/article/view/11/12

24. Blockly. https://developers.google.com/blockly

25. Hour of code. https://hourofcode.com/

26. Pattis, R.E.: Karel the Robot: Gentle Introduction to the Art of Programming with Pascal. Wiley, Hoboken (1981)

27. Vaníček, J., Šimandl, V.: Participants' perception of tasks in an informatics contest. In: Kori, K., Laanpere, M. (eds.) ISSEP 2020. LNCS, vol. 12518, pp. 55–65. Springer, Cham (2020). https://doi.org/10.1007/978-3-030-63212-0_5

Design and Analysis of a Disciplinary Computer Science Course for Pre-service Primary Teachers

Jean-Philippe Pellet[(✉)] [iD], Gabriel Parriaux[iD], and Morgane Chevalier[iD]

University of Teacher Education, Lausanne, Switzerland
{jean-philippe.pellet,gabriel.parriaux,morgane.chevalier}@hepl.ch

Abstract. According to new curricula being introduced in Switzerland, primary teachers have to teach concepts related to computer science (CS), but most of them have never been through a CS course themselves. At our university of teacher education, we have introduced a new disciplinary CS course for pre-service teachers, aiming to provide them with basic CS foundations to better grasp, contextualize, and explain the CS topics they will bring to their classrooms. This "experience report" paper describes the structure and design choices of the new disciplinary course. We propose a thematic split of the relevant topics to discuss and highlight strategies to make the course relevant for our audience. The declared and effective learning outcomes are then analyzed, topic by topic, through crossing survey responses and exam data. We also use survey data from a year later, polling the same participants again for relevance of their learnings in the disciplinary course after being in classrooms and conducting activities in CS. Through this, success points and improvement areas of the new course as well as changes to be made for the next occurrences are identified.

Keywords: Teacher education · Computer science education · Disciplinary course

1 Introduction and Context

In the French-speaking part of Switzerland, the K–12 curriculum for digital education was updated in 2021. Next to the already existing axes "media" and "ICT", a new "computer science" axis (CS) was introduced. The new curriculum was presented with the main objective of developing so-called *digital citizenship* and *digital culture* of pupils. Prior to this reform, political changes had brought the topic of digital education in general—and CS in particular—in the spotlight. There seemed to be a general preoccupation in the economic, scientific, and political spheres, relayed by the media, that children in our country would not receive sufficient education in the digital domain. In canton Vaud (one of the French-speaking Swiss cantons), digital education was elected as one of the main projects of the government and an ambitious pilot project was launched

A. Bollin and G. Futschek (Eds.): ISSEP 2022, LNCS 13488, pp. 125–137, 2022.
https://doi.org/10.1007/978-3-031-15851-3_11

to introduce the new curriculum of digital education and a teaching of CS in eleven schools starting from primary level [8].

Our university of teacher education has thus had to adapt its own curriculum to give primary pre-service teachers (hereafter referred to as "students") the competencies needed to teach the content mentioned in the new school curriculum, in particular CS. As those students had never had the opportunity to study CS in their school career, most of them needed education not only on the didactical side of CS, but first and foremost on its disciplinary aspects. Hence, a special course has been set up in our university of teacher education, focusing especially on disciplinary content of CS. Participation is optional, but all students have to go through the assessment at the end of the course. It is followed a semester later by a didactical course (not further discussed here).

In our educational system, universities of teacher education rarely provide disciplinary courses to pre-service teachers. For most of primary school disciplines, students' knowledge acquired during past school years is sufficient to follow the linked didactics courses (or else they join a regular university to study disciplinary content). Because of its novelty, and also to better understand what happened with teachers' knowledge in CS, we framed this new disciplinary teaching with a research setup capable of providing us with the information necessary for its regulation. This paper presents our setup and our main results.

Our main research questions are:

RQ1: How did students view CS before the course and has this view evolved?
RQ2: How did students self-assess their mastery of various subfields of CS before and after the course, and how accurate is this self-assessment?
RQ3: A year later, after trying out CS activities in the classroom, how did student retrospectively view this disciplinary new course?

This paper has the following structure: in Sect. 2, we mention analyses of similar initiatives or approaches to CS education of future teachers. In Sect. 3, we detail the structure of our new course and justify the design decisions. We talk about the source of the data we collected to answer our research questions in Sect. 4 and analyze it in Sect. 5. We conclude finally in Sect. 6.

2   Related Work

CS is relatively recent compared to other sciences, and this is even more true for its teaching. Only recently did it enter compulsory-school curricula in several countries (e.g., New Zealand, 2011 [2]; Estonia, 2012 [16]; the U.K., 2014 [15]; etc.). Most CS core knowledge is as new for the students as for their teachers. Despite this, teachers must be able to carry out a didactic transposition [4]. On the one hand, this professional gesture implies acquiring knowledge in CS (first level of transposition) and, on the other hand, reflecting/planning/proceeding to the transmission of this knowledge to their students (second level of transposition) [4]. In this study, we are interested in this first level of transposition on the part of the pre-service teachers and, as such, many researches have looked into the CS conception among teachers. For instance, Funke et al. [10] interviewed six primary-school teachers on their opinions towards CS courses at primary schools. Results unsurprisingly showed that teachers need to be trained to access the core CS concepts. This kind of need has also been reported in another similar study [13], in which educators could improve upon teachers' misconception about CS in only three training days. Besides, more and more recommendations to decision-makers in education systems encourage pre-service teachers to take CS courses as part of their teaching degree programs [5] to meet minimum content and knowledge requirements.

Moreover, a survey [7] among 116 secondary-school CS teachers about the integration of CS in primary education showed that essential topics for primary school should be introduced into primary-teacher training to ensure that they do not pass on their misconceptions to students. In this spirit, Repenning et al. [14] designed and experimented with a core CS course among pre-service primary teachers ($n = 600$). Results showed that teaching a mandatory CS class for pre-service teachers helps change the representations of these actors (particularly women) regarding CS and digital technologies. Nevertheless, results still reported a lack of confidence in implementing CS concepts in the classroom.

Another study [17] investigated the lack of confidence of in-service K–12 teachers concerning their self-efficacy in assessing Digital Technologies against the Australian Teacher Professional Standards and various assessment practices. Teachers reported that they need time and support to develop assessment strategies for this new area.

This raises the question of the perceived usefulness of CS knowledge to primary school teachers. Unfortunately, to our knowledge, no study reports the needs and feedback of in-service and pre-service teachers regarding the transfer of core CS knowledge that they have been able to carry out in class with their pupils.

As a result, given state of the art, it seems necessary to offer training to pre-service teachers on the core CS knowledge—ideally, spread out over time and of at least three days. Concrete links with society should be emphasised in such training to enable pre-service teachers to foresee a transfer of this knowledge and a didactic transposition appropriate to the maturity of their pupils.

3   Structure and Content of the New Disciplinary Course

The time slots obtained for this course encompassed 6 half days over the course of one semester, which is equivalent to Prieto et al.'s three full days [13].

3.1   Syllabus

The syllabus of the disciplinary course, according to our institution's policy, should be based on the high-school syllabus of the same topic. However, at the time of course preparation, there was no mandatory CS course in high schools yet;[1] our own syllabus was then based on circulating unofficial drafts.

---

[1] Such a course is actually due to be introduced in 2022–2023.

There is a general trend in CS education to move away from a coding-centric approach [1,9]. We thus wanted to ensure that we were not only focusing on programming and ended up with the following three main subfields: (a) data representation; (b) algorithms and programming; (c) machines and networks.

As mentioned in the introduction, mandatory CS in schools, in the view of our minister of education, should serve a "digital citizenship" goal rather than a mainly technical goal. According to this point of view, while the basis of data representation and programming should be taught, it is equally important that the societal implications linked to the usage of technology in the general public be exposed and discussed [11]. To embody this perspective in the new course, we discussed societal issues linked to the technical topics in each of the 6 sessions and made them an integral part of the syllabus.

We also strongly felt that we needed a common thread along the course. The bigger part of our audience has no special appetency for technical matters: we could not just place next to each other a series of themes deemed relevant by us without a strong, visible link between them. We thus picked *web search* as a common thread, and arranged the conceptual topics of the syllabus around the exploration of what really happens at various stages of running a web search.

Here is the final six-session syllabus, formulated in terms of the common thread and linked to societal issues:

1. **Data representation**. *"Computers work with 1s and 0s. When you do a web search, the search terms are also 1s and 0s—so are the results you get, be them text or images. Let's find out how we can represent such data with bits."* ⇒ Binary representation of positive integers; representation of text, basics of bitmapped images. *Societal issues:* Energy consumption of storage systems and large communication infrastructure in response to growing usage.

2. **Computer architecture**. *"We now know how our web request will be represented. Let's now look at how these 1s and 0s travel through the electronics inside a computer and how that electronics can be build to process that information."* ⇒ Basic logic gates; high-level view of components such as CPU, storage devices, and sensors. *Societal issues:* History of CS and automated machines; influence of war-time goals (deciphering, ballistic computations) on the development of computers.

3. **Network and cryptography**. *"After exiting our computer, our web search goes through the internet to reach the search provider's servers. How it is relayed by the intermediaries involved? How can I prevent these relays from reading what's in my request?"* ⇒ Packet switching, basic routing and idea of protocol; examples of symmetric ciphers, common attacks, and principles of asymmetric cryptography. *Societal issues:* Disparities in the world for internet access; pros and cons of strong ciphers and end-to-end encryption.

4. **Programming I**. *"Our request has reached the remote server. To be answered, it is processed by the searched company's software. What is software and how do you instruct a machine what to do? Through programming."* ⇒ Using Python's `turtle` module: simple movement, simple loops, simple func-

tion definitions, without or with one parameter. No variables at this point. *Societal issues:* Open source/free software, licensing (not limited to software).

5. **Programming II.** *"The previous examples have showed us how to give instructions to a computer; this session will make more explicit the way data is referenced and handled in programming languages through variables that can represent values that are still unknown when the software is written."* Same programming environment: variables, `if` statements and conditions with variables, simple lists (definition and iteration). *Societal issues:* Data collection, profiling, recommendation algorithms, and third-party cookies.

6. **More algorithms and AI.** *"CS is more than web searches. Let's examine graphs, which allow us to model many problems, and a related algorithm, which is used in our GPS but not only there. Finally, let's talk about AI: what it is, what it isn't, and a little bit of how it works."* ⇒ Without programming: concept of graph, tracing of Dijkstra's algorithm, applications. AI: high-level principles of a rule-based classification algorithm. *Societal issues:* Importance of training data for AI systems and awareness of bias-reproducing systems.

There were many other topics (technical or societal) we had deemed worthy of interest that did not end up making it into the syllabus for time constraints. Many steps in the web-search-processing story are still missing. The goal of the common thread is to arrange the selected topics in a tractable sequence rather than provide a full explanation of the chosen phenomenon.

3.2 Operational Planning

The course ended up being given entirely remotely due to COVID-19. We prepared 4 to 6 videos of between 5 and 15 min for each of our 6 sessions, for a total never surpassing 60 min. Our aim was to keep the video time at a maximum of 60% of what the actual lecture time would have been.

We made a creative use of the Label element in Moodle to structure the subsections, link each of them with a short list of expected learning outcomes, and added to each of them practical exercises, the solutions to which were available to students and explained in details, sometimes with more videos.

Societal issues cannot readily be linked to practical exercises. In order not to limit ourselves to videos on these issues, questions were asked at the end of each videos, in a "food for thought" way—with no correct or incorrect answer. Arguments serving these discussions were provided in the "solutions" part.

In addition to the material on the Moodle page, we opened for each session a Zoom room for 120 min, creating several breakout rooms which corresponded to the session's subsections. This always included a room to discuss the open questions related to the societal issues. Each breakout room was staffed with one instructor and the students could thus freely navigate between them according to the concepts they were stuck with or wanted to discuss.

As a last means of interactions, a traditional Moodle forum was made available for public, asynchronous questions and answers about the course.

3.3 Evaluation

Evaluation was done during an open-book 90 min Moodle quiz comprising 28 multiple-choice (MC) questions (which were corrected automatically) and 4 open-text questions (which were corrected manually).

The MC questions were "rich" in the sense that they were not only text, but embedded images and (with the help of HTML `iframe`s) interactive logic diagram or code editors. The open-text questions asked the students to describe, with a few sentences of their own writing, their understanding of the societal issues discussed in the course and short analysis of a small related example situation. Typically, such topics cannot adequately be assessed with MC questions.

We insisted on the open-book policy, convinced that, especially in the field of CS, any evaluation requiring students to learn certain things by heart was assessing capabilities related to the lower levels of the cognitive domains of Bloom's taxonomy (Comprehension and Understanding), and we were trying as much as possible to focus on the higher levels (more specifically, Application and Analysis for technical topics, and Analysis and Evaluation for societal issues).

4 Data Collection and Methodology

To answer our research questions, we used data from the following sources:

- A survey given before the beginning of the new course, which contained question about age, gender, previous education, and included a section meant to capture their representation of CS and their declared a priori mastery of the main subtopics of the course (see details below);
- A post-course survey, with the same questions about their representation of CS and their declared a posteriori mastery of the same subtopics;
- Grades from the examination described above, given at the end of the course, with a detailed split of the points among the same subtopics whose declared mastery was asked about in the surveys;
- A survey given a year later to the same students, asking if they viewed what they had learned in the new course as useful for the following didactics course and for the classroom activities they had conducted in the meantime.

The grade records were anonymized and uniquely identified by some Moodle-generated ID. The same ID was automatically filled out in the pre- and post-course surveys, enabling us to automatically link the survey responses while keeping them anonymous.

4.1 Common Pre- and Post-Course Survey Questions

To address *RQ1*, both pre- and post-course surveys included questions to depict the students' view of CS before and after the course. They were asked to rate these statements on a Likert scale ranging from 1 (disagree completely) to 6 (agree completely): "To me, computer science..."[2] [12]:

---

[2] Students were also asked to answer an free-text question on how they would describe CS. Size constraints do not allow the inclusion of the analysis of those results here.

1. is mainly applied mathematics
2. does not really have permanent components and is constantly evolving
3. changes rapidly but rests on stable notions
4. has theoretical foundations
5. is mainly about learning how to use office software
6. is primarily about practical knowledge rather than concepts and notions
7. is the major science of the 21st century

To address $RQ2$, students were also asked to rate their mastery of the following subtopics on a scale ranging from 1 (no mastery) to 5 (excellent mastery): 1. binary data representation; 2. CPU and computer architecture; 3. cryptography; 4. programming; 5. algorithms and AI; and 6. computer networks.

The survey enabled us to observe how each student changed their view of CS and how their declared mastery evolved. Moreover, we could determine the correlation between the declared post-course mastery and the actual exam results for first 5 subfields listed above (subfield 6 was made optional and was absent from the final exam).

4.2 Year-After Survey

In the year after the disciplinary course, student have followed a didactics course and have had the opportunity to conduct a CS-related activity in a classroom. Since the new course was supposed to provide the foundations for the didactics course, we were interested in asking students' opinion through these two questions to address $RQ3$: (a) How useful did you find the disciplinary course for the activity you conducted? (rated on a 7-point Likert scale), and (b) Retrospectively, how adequate did you find the level of the disciplinary course? (rated on a 5-point scale: way too hard/too hard/adequate/too easy/way too easy).

We were especially interested in the year-after (rather than the right-after) opinion since it would be difficult for students to evaluate the adequacy of such a course without practical experience with actual classroom activities.

5   Analysis and Discussion

357 students were registered for the course and obtained a grade. Out of them, 284 filled the pre-course survey; 130 filled the post-course survey (114 filled them both); and 117 filled the year-after survey. The Demographics subsection thus rests on the 284-sample dataset; the analysis covering pre/post comparisons and the year-after opinion use the 114- and 117-sample datasets, respectively.

5.1   Demographics

About 65% of the respondents are between 18 and 22 years old; 22% are between 23 and 30; 8% between 31 and 40, and the rest 5% are older. 85% are female.

The highest degree of 85% of the respondents is a high-school degree. About 12% have a college degree; about 3% have another degree (professional or other).

87% of all students passed the exam on the first attempt.

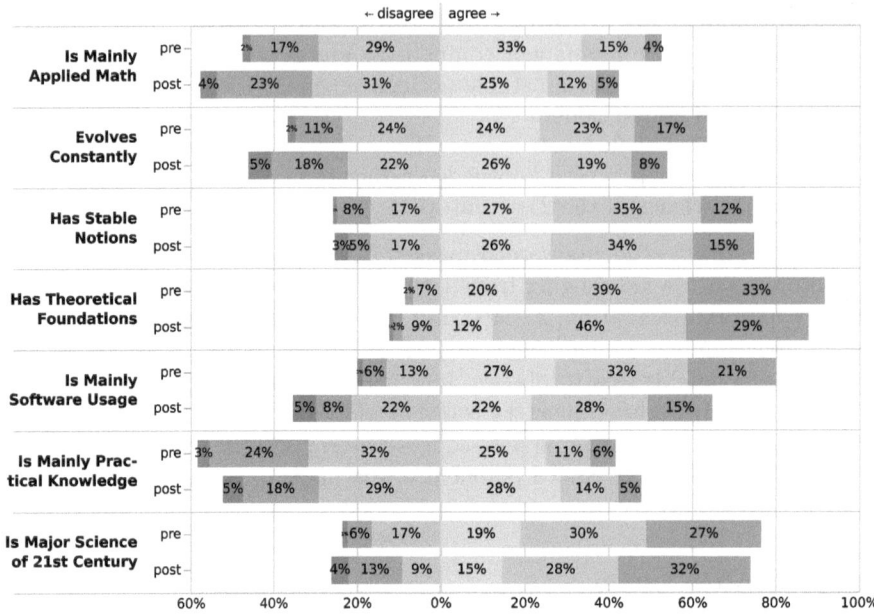

Fig. 1. Shift of opinions on what CS is according to the students, following the 7 questions described in Sect. 4.1, rated on a 6-level Likert scale. Red bars indicate (center-to-left) "somewhat disagree", "disagree", and "strongly disagree"; green bars indicate (center-to-right) "somewhat agree", "agree", "strongly agree". (Color figure online)

5.2   Representation of Computer Science

We now analyze students' views of CS through their compared (pre- and post-course) opinion on the 6 assertions listed in Sect. 4.1, shown on Fig. 1.

Students are almost evenly split on whether CS is mainly applied mathematics. After the course, they tend to reject this assertion more (even if the shift is on the verge of being significant at the .05 threshold: Mann–Whitney's $U = 20568$, $p = .0537$). The diversity of the subtopics and the discussion of the societal issues were meant to favor such a shift and a conceptual separation of math and CS.

Although CS is still viewed by the majority as constantly evolving with no permanent components, there is a significant shift ($U = 21749$, $p = .0029$) occurring. In hindsight, we should have phrased this question differently, as it mixes two dimensions (having permanent components and evolving constantly, both of which can be argued to be true) into a single statement.

More than 75% of respondents agree that CS has stable notions (this does not change between pre and post). Even more agree that CS has theoretical foundations. We were surprised to see more students disagreeing on this after the course. Even though the difference is not statistically significant, we ideally would have liked to see the opposite shift: we believe this is due to our very practical approach which tried to include as little theory (and as little math) as possible.

We were pleased to see significantly more disagreement on CS being mainly software usage ($U = 21507$, $p = .0055$), even though a majority still agrees. There was no significant shift on the two remaining questions of our surveys.

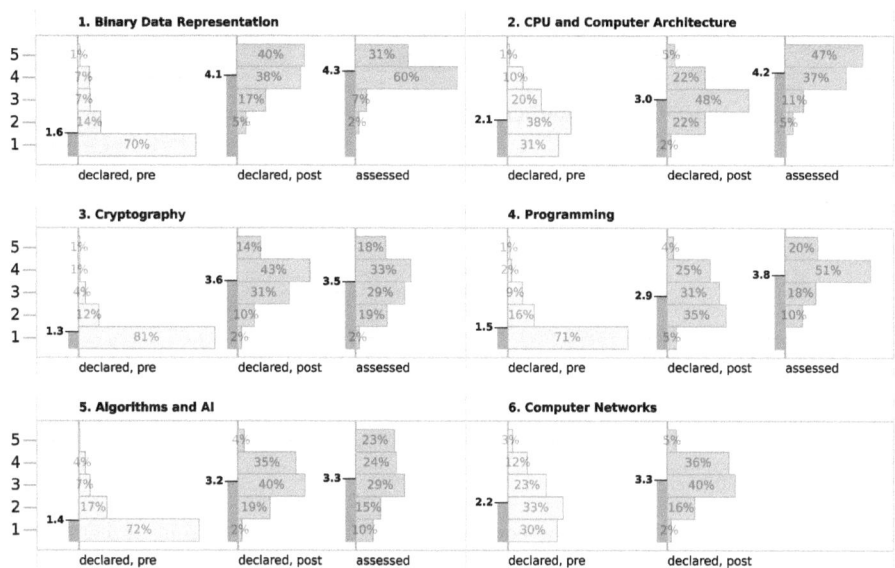

Fig. 2. Declared pre- and post-course mastery of CS subfields taken from surveys, compared to the performance on the final exam.

5.3 Declared and Assessed Mastery Levels of Subtopics

The results shown on Fig. 2 show, for each of the 6 subfields, a distribution of declared mastery on a scale ranging from 1 (no mastery) to 5 (excellent mastery). Pre-course data is shown first (yellow), then post-course data (green), and last (blue), we show the scores obtained by averaging over the grades of the relevant questions in the final exam and rescaling to reach the same 1-to-5 range. Subtopic 6 was made optional and no exam question was asked on it. The vertical grey bars show the mean of each distribution.

The declared pre-course level was almost the same for subtopics 1, 3, and 5, and the means of the declared post-course levels are very close to the final grades. Some students slightly overestimated their understanding of binary data representation and slightly underestimated that of algorithms of AI. On the two subtopics 2 and 4, students significantly underestimated their understanding with respect to our exam questions. These are the two topics where our course exercises included synthesis questions—small logic circuits in an interactive tool for subtopic 2 and short Python programs with `turtle` for subtopic 4. These were experienced as difficult by students. On our final exam, these subtopics were addressed with multiple-choice question, which definitely made them easier to achieve than if they had also been synthesis questions.

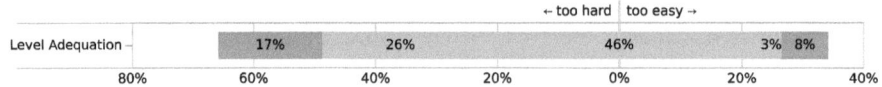

Fig. 3. Perceived adequation of the level of the course a year later.

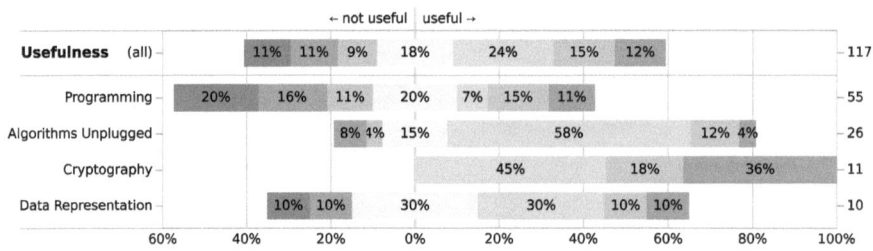

Fig. 4. Perceived usefulness of the course, a year later, split according to conducted classroom activity.

5.4   Year-After Opinion

A year later, 43% of the students who expressed their opinion said the course had been too hard, 11% too easy, and 46% adequate, as shown in Fig. 3. The imbalance between the two extremes has since then made us reconsider the inclusion of some more involved notions, especially in programming (Python's lists) and algorithms (some properties of graphs).

About half of the students found the disciplinary course had been useful to them (Fig. 4, first row). We were nevertheless surprised that almost a third said it had not been useful, so we split the analysis according to the type of activity the students had conducted in the classroom (filtering out activity types with fewer than 5 instances). The next rows in the same figure show the significant differences between them. The negative ratings mostly come from students who have conducted programming activities, while those having dealt with cryptography-related activities had unanimously found the course useful.

The main difference between these two subtopics is how closely what we do in the new course is related to what they can actually conduct as activity in the classroom. The programming activities are quite different: the course uses Python, but they will use robots or Scratch Jr in classrooms—which is quite different, even though underlying concepts may coincide. For cryptography, we begin with Caesar's cipher and, although we also discuss more advanced polyalphabetic ciphers and several attack types, they can directly conduct a classroom activity based on Caesar's cipher, making the link immediately clear.

While research shows that being correctly educated about the disciplinary content is crucial for teachers, it also highlights the fact that it is not necessary to go far beyond the level of corresponding knowledge that they teach in their class. The more one goes beyond a minimal base of disciplinary knowledge as taught at a given level, the less added value this disciplinary knowledge brings to teaching [6]. But we also know from Bruner [3] that a knowledge of individual concepts is not sufficient. There must be an understanding of the way concepts are organized together, of the underlying principles that support them.

Let us be reminded that a subsequent mandatory didactics course exists to precisely present classroom activities. We have no clear way of actually knowing that the basics of programming they acquired in the new course did not make them more effective programming teachers. But seeing how closeness to classroom activities is beneficial to perceived usefulness has made us change the design of future occurrences of the course so as to always start with a motivating example very closely linked to an activity they will be able to conduct.

6   Conclusion

We have described the primary-teacher-education context in which we deemed necessary to introduce a new introductory course to CS. We have outlined its specificity and its syllabus, highlighting our common-theme approach and the discussion of the broader societal issues. We explained the all-remote modalities related to the COVID-19 situation.

We have analyzed data from pre- and post-course surveys, from the final exam, and from a year-later survey, after the students had conducted CS activities in actual classroom. Data shows that they view CS differently on a subset of statements on which we asked them to express agreement; notably, that CS is not about software usage and does have permanent components, although it evolves constantly. We noted that our theory- and math-poor approach has not reinforced much the impression that CS has theoretical foundations, although the course was also meant to convey the idea that CS is science indeed.

Declared mastery and exam questions show that topics treated with exercises involving creating programs or small logic circuit (where synthesis is needed) reduce the perceived mastery compared to other topics like data representation and cryptography (where exercises rather test understanding and analysis than synthesis).

Finally, the year-after survey showed greatly different perceived usefulness of the new course depending on the type of conducted classroom activity, even though all such activities were conceptually linked to the course. Perceived usefulness was maximal when the course not only conceptually coincided with the conducted activity, but also directly treated (and expanded on) scenarios that could form a direct basis for that activity.

These results have enabled us to make data-driven adjustment to the new course so as to better highlight some fundamental aspects of CS as well as increase the perceived usefulness of the course.

References

1. Astrachan, O., Briggs, A.: The CS principles project. ACM Inroads **3**(2), 38–42 (2012)
2. Bell, T., Andreae, P., Robins, A.: A case study of the introduction of computer science in NZ schools. ACM Trans. Comput. Educ. (TOCE) **14**(2), 1–31 (2014)
3. Bruner, J.S.: The Process of Education. Harvard University Press, Cambridge (2009)
4. Chevallard, Y.: On didactic transposition theory: some introductory notes. In: Proceedings of the International Symposium on Selected Domains of Research and Development in Mathematics Education. pp. 51–62. Comenius University Bratislava, Czechoslovakia (1989)
5. Computer science teachers association, code.org advocacy coalition: state of computer science education (2018). https://code.org/files/2018_state_of_cs.pdf
6. Darling-Hammond, L.: Teacher quality and student achievement. Educ. Policy Anal. Arch. **8**, 1 (2000)
7. Dengel, A.: Opinions of CS teachers in secondary school education about CS in primary school education. In: Proceedings of the 12th Workshop on Primary and Secondary Computing Education, pp. 97–98 (2017)
8. El-Hamamsy, L., et al.: A computer science and robotics integration model for primary school: evaluation of a large-scale in-service K-4 teacher-training program. Educ. Inf. Technol. **26**(3), 2445–2475 (2020). https://doi.org/10.1007/s10639-020-10355-5
9. Fincher, S.A., Robins, A.V.: The Cambridge Handbook of Computing Education Research. Cambridge University Press, Cambridge (2019)
10. Funke, A., Geldreich, K., Hubwieser, P.: Primary school teachers' opinions about early computer science education. In: Proceedings of the 16th Koli Calling International Conference on Computing Education Research, pp. 135–139 (2016)
11. Paoletti, F.: Épistémologie et technologie de l'informatique. Revue de l'Enseignement Public et Informatique **71**, 175–182 (1993)
12. Parriaux, G., Pellet, J.-P.: Computer science in the eyes of its teachers in French-speaking Switzerland. In: Brodnik, A., Tort, F. (eds.) ISSEP 2016. LNCS, vol. 9973, pp. 179–190. Springer, Cham (2016). https://doi.org/10.1007/978-3-319-46747-4_15
13. Prieto-Rodriguez, E., Berretta, R.: Digital technology teachers' perceptions of computer science: it is not all about programming. In: 2014 IEEE Frontiers in Education Conference (FIE) Proceedings, pp. 1–5. IEEE (2014)
14. Repenning, A., Lamprou, A., Petralito, S., Basawapatna, A.: Making computer science education mandatory: exploring a demographic shift in Switzerland. In: Proceedings of the 2019 ACM Conference on Innovation and Technology in Computer Science Education, pp. 422–428 (2019)
15. Sentance, S., Csizmadia, A.: Computing in the curriculum: challenges and strategies from a teacher's perspective. Educ. Inf. Technol. **22**(2), 469–495 (2017)

16. Shin, S., Bae, Y.: Study on the implications about curriculum design through the analysis of software education policy in Estonia. J. Korean Assoc. Inf. Educ. **19**(3), 361–372 (2015)
17. Vivian, R., Falkner, K.: A survey of Australian teachers' self-efficacy and assessment approaches for the K-12 digital technologies curriculum. In: Proceedings of the 13th Workshop in Primary and Secondary Computing Education, pp. 1–10 (2018)

Textbooks and Materials for Teaching Computer Science in Slovenia

Špela Cerar[1]([✉])[iD], Matija Lokar[2][iD], Gregor Anželj[3][iD], Andrej Brodnik[4][iD], and Irena Nančovska Šerbec[1][iD]

[1] Faculty of Education, University of Ljubljana, Ljubljana, Slovenia
{spela.cerar,irena.nancovska}@pef.uni-lj.si
[2] Faculty of Mathematics and Physics, University of Ljubljana, Ljubljana, Slovenia
matija.lokar@fmf.uni-lj.si
[3] Gimnazija Bežigrad, Ljubljana, Slovenia
gregor.anzelj@gimb.org
[4] Faculty of Computer and Information Science, University of Ljubljana, Ljubljana, Slovenia
andrej.brodnik@fri.uni-lj.si

Abstract. In Slovenia, the RINOS working group has proposed changes to the curriculum and the introduction of computer science (CS) as a compulsory subject in accordance with the K12CS framework. An important step in the introduction of the subject is the development of new teaching materials and environments for teaching CS. Currently, most Slovenian materials for teaching CS are collected in the *Lusy library*. These include e-textbooks, materials for teaching CS, booklets of tasks and solutions, systems for various CS competitions, and platforms supporting practice of programming. Good teaching materials for CS should be relevant, interactive, easy to navigate, and promote active learning. Because CS education is a relatively new field, its didactics is almost constantly improving. Therefore, it is necessary to prepare teaching materials that can be easily modified, adapted and used for active learning. Various teaching materials with these properties can be found in the *Lusy library*. We conducted a qualitative survey among teachers of CS to assess the usefulness of existing materials and environments for teaching selected thematic units and to gain insight into teachers' needs. Our findings will provide the basis for developing useful practical teaching materials and activities for a future compulsory subject.

Keywords: Teaching materials · I-Textbooks · Interactive learning materials · Learning environments · Assessment platforms

1 Introduction

The first e-textbooks were digital copies of printed textbooks with easier navigation via hyperlinks and the ability to search the content [21]. Nowadays however,

A. Bollin and G. Futschek (Eds.): ISSEP 2022, LNCS 13488, pp. 138–149, 2022.
https://doi.org/10.1007/978-3-031-15851-3_12

a good e-textbook differs significantly from a printed textbook. The digital format makes it easier to update the content, allows the use of multimedia and interactivity, and enables customisation to an individual or target a group of learners [15]. When creating e-textbooks, the advantages of their digital format must be properly exploited. The quality of an e-textbook is measured by how it contributes to the understanding of the topic presented. Technological advances have enabled new pedagogical approach. The ease of updating and customization is especially useful in the field of CS, because of the rapid development of CS curricula and its didactics. It is also useful that e-textbooks provide support for interactive elements, interactive tasks with instant feedback, and enable various analytics. Such e-textbooks are often referred to as i-textbooks [18].

Today, we can use various tools to develop i-textbooks and materials, such as free platforms openDSA [9] or Runestone, or payable platform zyBook [16]. These platforms not only support the creation of new materials, but also include examples of learning materials and a set of tasks with immediate feedback. Use of these platforms, makes it possible to use images, assess the knowledge through closed-ended questions, animate the implementation of the program code, write the code and check its correctness, etc. Studies have provided evidence that i-textbooks improve student performance and engagement compared to traditional textbooks [5,6]. Also a recent study by a research group at Aalto University found that the use of i-textbooks improved student motivation and learning compared to the use of static e-textbooks [19].

The development of textbooks and materials is influenced not only by the didactics of the subject, but also by the position of the subject in the curriculum. Currently, in Slovenia, CS is taught in basic education as an elective subject in grades 4–9. In the gymnasium it is taught as a compulsory subject in the first year, and as an elective matura subject in subsequent years. The current curricula are written openly giving teachers the freedom in teaching the subject and the selection of topics. On one hand, this freedom is beneficial, but on the other, it introduces problems. For example, the basic education curriculum lists many learning objectives without specifying which ones are fundamental and must be achieved by every student. As a result, teachers choose learning objectives according to their own preferences, which, unfortunately, most frequently only contain digital literacy.

In the elective CS courses in grades 4–6, the groups may consist of students of different ages, from different grades, and with different levels of knowledge, which makes the teaching even more challenging. In grades 7–9, the curriculum is still the same as in 2002 and is mainly based on digital literacy.

We are currently in the phase of introducing a new compulsory subject CS throughout the educational vertical. Namely in 2017, the Ministry of Education, Science and Sport of Slovenia set up an expert group RINOS. In Slovene RINOS stands for *Strokovna delovna skupina za analizo prisotnosti vsebin Računalništva in INformatike v programih Osnovnih in Srednjih šol ter za pripravo študije o možnih spremembah*, which translates to English as an *Expert working group for analysis of Computer Science topics presence in primary and secondary schools, and for a preparation of proposal of possible changes.* In 2018 [3], the RINOS Expert Group made a proposal, that was also approved by the The Slovenian Academy of Sciences and Arts, to change the curricula by introduction of the compulsory subject CS into primary and secondary schools. Following the K12CS framework, the proposed curriculum covers the following topics: data and analysis, algorithms and programming, computer systems, networks and the Internet, and the impacts of computing [12]. However, shifting the focus from digital literacy to core CS concepts requires a change in teaching methods and the creation of appropriate instructional materials.

In the rest of the paper we introduce some textbooks and materials for teaching CS in Slovenia, which is followed by the presentation of survey results on the use of and need for textbooks and materials for learning and teaching CS. It was conducted among CS teachers in Slovenian primary and secondary schools. Finally, we summarise the findings from the literature and the research mentioned above.

2   Textbooks and Materials for Computer Science Teaching

In recent years, numerous e-materials and e-textbooks for CS education have also been created in Slovenia. Most of them can be found at the *Lusy library*[1] (Fig. 1).

2.1   Materials and Platforms

In this section we present some of the most popular and widely used online materials used by Slovenian teachers.

Vidra - CS Unplugged. The portal *Vidra* contains a collection of free learning materials for learning CS with games and puzzles, using cards, strings, crayons, and lots of movement. The Slovenian adaptation is based on the well-known CS Unplugged [2]. The materials provide CS teachers with interesting ideas and descriptions of activities. They can also be used in other subjects, especially mathematics, in clubs and as extracurricular activities. The materials cover topics such as binary numbers, graphs, sorting algorithms, and artificial intelligence [4].

Code.org. This is a website for learning CS fundamentals and includes online and hands-on activities. It teaches students computational thinking skills, problem solving, programming concepts, and digital citizenship. The developers have created learning paths that allow students to solve problems on their own. Substantial part of Code.org is translated into Slovene and is used by teachers mostly as additional material in class.

---

[1] https://lusy.fri.uni-lj.si/ucbenik/.

Fig. 1. *Lusy library*, central access point for i-textbooks and e-materials for CS.

Bober. The ACM Slovenia Bober Competition (from here on Bober) is a part of the International Challenge on Informatics and Computational Thinking Bebras. It is aimed at students from the 2nd grade of primary school to the end of secondary school. A very important result of this activity is the booklet with tasks and their solutions, published annually under the CC BY-SA license. It serves as a teaching material widely used to promote development of computational thinking and as a preparation for the competition. The booklet contains tasks and correct solutions with a detailed explanation of the correct solutions. For each task there is also a description of the CS background.

Pišek. The Pišek platform supports learning introductory block-based programming. It was created in 2018 based on the French system Algorea [10]. The portal contains a collection of tasks to be solved using the programming language Blockly and gives the user immediate feedback on the correctness of the solution. There are three main types of tasks in Pišek: tasks on the grid, tasks with turtle graphics, and "classic input/output" tasks. Tasks have different approaches to solutions: assembling a program, correcting a program, completing the program, and Parsons-type tasks. The platform is used on a national level for the ACM Slovenia Pišek competition, which annually wraps up with a booklet of tasks and their solutions.

Project Tomo. Project Tomo provides a rich library of tasks. When solving tasks, Tomo provides students with an immediate feedback on the correctness of the code. Teachers can set up their own "classrooms" with a selection of programming tasks for their students. They can use tasks from a library or create their own. Practical experience of high school teachers show that Tomo is particularly useful when teaching groups of students with heterogeneous knowledge [11,14]. Teachers pointed out that the system helps them prepare learning materials, monitor students' progress, and analyse their work. It facilitates teachers to individualise instruction and enable students to progress faster.

Putka. The set of platforms concludes Putka that is used predominantly for competitive programming on a national (ACM Slovenia RTK - national level Olympiad in informatics, ACM Slovenia UPM - national level of ICPC competition) and international level (CEOI, CERC). It also contains a rich collection of tasks that is accompanied with their solutions and explanations in *Zbirka rešenih nalog* (Booklet of solved tasks).

2.2 I-Textbooks

Due to changes in the curricula in 2013, it was necessary to write a new textbook for Informatics that replaced the old one, which was oriented very strongly towards computing literacy. Authors decided to write an interactive textbook that would be accessible online free of charge [17]. The i-textbook Informatics 1 covers four topics: programming and algorithms, systems, networks and distributed systems, and informatics and society [1]. It is well received by teachers, especially since it is constantly updated, mostly through teachers' input (e.g. errors discovered, suggestions). At the moment, the i-textbook Informatics 2 is also being developed, covering topics of information presentation, knowledge technology, and object-oriented programming. The authors plan to give teachers even greater opportunities to actively participate in the creation of the i-textbook, especially by providing opinions and evaluations of the content itself.

I-textbook is designed so that one can create different sequences of learning units and add own learning units as needed. The modularity and the possibility to adapt the i-textbook to one's own needs are considered a great advantage.

Since the i-textbook can be used also offline, it is designed as a series of interconnected static web pages. Interactivity is implemented using JavaScript, which allows for a single web page to be executed locally in a browser. In addition to the content, the authors also focused on the use of i-textbooks on different devices. This was mainly achieved by automatically adapting most of the content to the screen size.

On the other hand, when the i-textbook is used online, it provides a rich set of hyperlinks to other materials and platforms. For the former the inclusion of appropriate matura examination tasks with each learning unit is being implemented. Each learning unit is linked to the national SIO.si platform with class materials, and to the Tomo platform.

The chapter on programming and algorithms was reused in the i-textbooks *Slikovno programiranje* about block-based programming and *Malina in piton* on physical computing. The first uses Blockly instead of Python as the programming language and links to Pišek instead of Tomo. The second i-textbook uses the physical computing as a means to learn programming.

All of the available i-textbooks are released under Creative Commons license, more specifically they are released under CC BY-NC-SA 2.5 SI.

The usefulness of Informatics 1 i-textbook was evaluated with a survey of 61 CS teachers [17]. More than 90% of the teachers felt that the content was well explained. The interactive elements were also highly rated, especially the animations, the integrated Python interpreter, and the tasks with immediate feedback. Figure 2 shows the most frequent uses of the textbook.

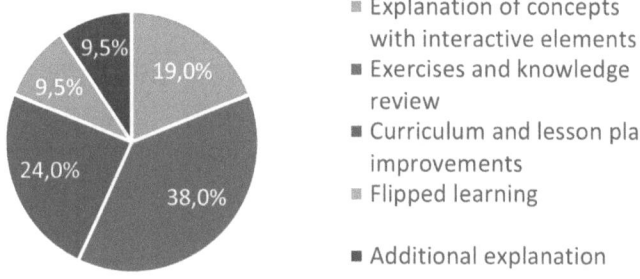

Fig. 2. Frequency of different uses of i-textbook in lessons.

3 Research on Teacher's Opinions on Existing E-Materials and I-Textbooks for CS

Through a survey of CS teachers, we aimed to assess the usefulness of existing materials and environments for teaching selected topics. The study was based on a combination of qualitative and quantitative research methods. As part of the survey, we were interested in what materials teachers use, how often they use them, and in which grades. We wanted to know what their needs were for materials on various topics: computer systems, data and analysis, algorithms and programming, networks and the Internet, and the effects of computing. We also investigated what features of e-materials are most important to teachers, what their experiences have been with using them, and what requirements they have for the materials.

3.1 Sample Description and Analysis of Responses

CS teachers were invited to complete the survey in early 2022. Responding 156 teachers were of different age groups and have been teaching CS and informatics subjects for different periods of time. All age groups were well represented, with the majority of teachers (83%) having up to 25 years of teaching experience. The majority of teachers surveyed teach or taught various CS subjects at basic education. Computer club was run by 53% of the teachers surveyed. High school subject Informatics was taught by 29% of the teachers. Computer science subjects in vocational high school were taught by 16% of the respondents.

The sample of teachers surveyed seems to be very motivated to improve their teaching: 57% attend PD programs more than once per year and 28% attend PD at least once a year. 76% of respondents are mentors to students in the computational thinking competition Bober, 32% mentor students in the ACM Slovenia Pišek competition and 11% mentor students in the high school programming competition ACM Slovenia RTK. 19% of respondents are not mentors to students in CS competitions.

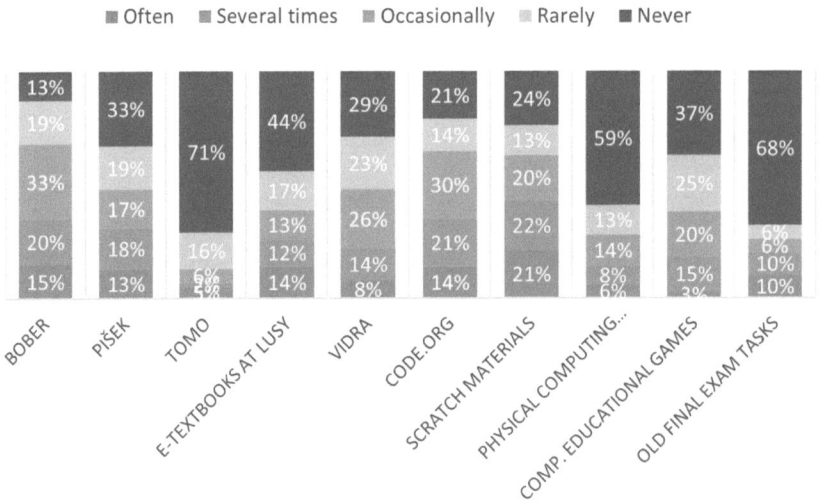

Fig. 3. Frequency of use of existing materials in lessons.

Teachers use a variety of materials in their teaching (Fig. 3), but most frequently they use tasks from the Bober competition. They most often use materials for learning Scratch, which is likely because basic education teachers commonly use Scratch in their classes, as it can be derived from their responses. Materials from Code.org, and Vidra are also popular. Teachers also use the Pišek portal, i-textbooks from the *Lusy library*, and educational computer games. Old matura tasks and the Project Tomo system are used by majority of high school teachers. Those teaching physical computing, use special learning materials for this purpose.

For practicing, teachers mostly use tasks from the Bober competition (62%), Scratch materials (61%), materials from Code.org (57%), and the Pišek portal (48%). Secondly, for frontal teaching they use Scratch materials (45%), materials from the portal Vidra (40%), i-textbooks from *Lusy library* (40%), tasks from the Bober competition (34%), and the portal Pišek (32%). Finally, as additional teaching materials, teachers use tasks from the Bober competition (45%), Scratch materials (27%), materials from Code.org (27%), portal Pišek (25%), materials from the Vidra portal (24%), i-textbooks on *Lusy library* (24%), and educational computer games (23%).

We also surveyed teachers on their needs and current situation of instructional materials. The teachers' answers show that Algorithms and Data Structures is the most well covered topic and that there are enough high-quality materials for this topic (29% of respondents agree). The greatest need for new high-quality learning materials is on the topic Networks and the Internet. The biggest gap between the need and the current situation is in the topics of Computer Systems and Impact of Computing. There is also a need to create additional high-quality materials for Data Science.

CS teachers stress the importance of integrating interactive tasks into i-textbooks (88%) and aligning content with the curriculum (82%). Three quarters of respondents deem automatic verification of task solution correctness an important part of i-textbooks. 65% of respondents indicate that ease of navigation in the e-textbook and inclusion of multimedia elements are also important. Nearly half (45%) of respondents would like e-textbooks to be customizable, to have content added to them, or to have certain parts removed or changed.

The question about experiences with using existing e-textbooks and interactive materials was answered by 67 primary education teachers, 30 high school teachers, and 10 vocational high school teachers. The opinions of primary education teachers are divided: 23 of them had good experiences, 17 had mixed experiences, and 11 had bad experiences. As an example of a good resource, teachers highlighted the portal Pišek and the interactive tasks on Code.org.

A very important comment was the observation that a number of high-quality interactive materials have been produced through different projects in the past, but these are no longer maintained and are consequently outdated. They also point out that many learning materials are not adapted to the basic education level and are therefore unsuitable for students at this age. The problems are non-systematic escalation of the complexity of tasks, and for students in grades 4–6 the lack of learning materials in Slovene. Teachers noted that there are many high-quality teaching materials in foreign languages. Many of them translate the resources, but due to an uncoordinated approach, their efforts are unnecessarily multiplied.

Most high school teachers had good (14 teachers) or mixed (13 teachers) experiences with e-textbooks and interactive learning materials. Only one teacher expressed the opinion that the available materials were inadequate and that she has to create them herself. The most frequently used resource for high school teachers is the i-textbook Informatics 1. It is perceived by teachers as useful, professionally relevant and also interesting for students. They also emphasise the importance of interactive tasks with immediate feedback.

The teachers in the secondary CS schools had good experiences with the use of e-textbooks and interactive learning materials. Only one teacher had a decidedly bad experience. As a disadvantage, they pointed out that e-textbooks are sometimes too in-depth for the level of knowledge they require from their students. The main positives they expressed are ease of use, and the interactive tasks that allow students to get real-time feedback.

Teachers of other secondary schools pointed out that there is a good i-textbook for high schools, but not for other secondary schools. A significant number of them use at least part of this i-textbook for their work. They miss a list of different learning materials to help them design their own learning materials for the students. They also pointed out the need for regular updates of i-textbooks and other learning materials.

Teachers pointed out the following reasons for integrating interactive learning materials into CS lessons:

- "Interactive materials are more attractive to students because they provide a sense of involvement."
- "E-textbooks are very welcome in classroom, because they allow the teacher to let students work at their own pace."
- "In high school, the e-textbook is excellent and students get the full support they need in the subject of informatics. The students themselves were very complimentary about the i-textbook in the surveys we conducted at the end of the school year."

At the same time, teachers emphasize the importance of direct contact between students and teachers and are not "afraid" that modern i-textbooks and resources will make the teacher's job obsolete. E-textbooks and interactive materials are useful as supporting material for learning new topics, consolidating knowledge, providing immediate feedback on solutions of various tasks, and formative knowledge assessment.

Teachers were also asked what they miss in existing e-textbooks and interactive teaching materials. Basic education teachers pointed out that many of the resources are not updated (10 teachers). Teachers would like to see a collection of materials (9 teachers) to assist them in teaching CS subjects in basic education. They also wish to have exemplary lessons and useful advice for teaching individual content areas (7 teachers). They would like to see high-quality interactive learning materials with tasks that are adapted to the knowledge and interests of basic education children (8 teachers) and allow for differentiation.

Among high school teachers, we see a slightly lower demand for new interactive materials, which might be related to the fact that the i-textbook Informatics 1 is available. However, teachers pointed out that not all learning objectives are covered (15 teachers). 9 teachers would like to see more tasks or collections of tasks, and would also like the tasks to allow for differentiation between students (3 teachers) and for the tasks to be more practical (2 teachers).

4 Conclusions and Plans

According to [8,15,20] the desired characteristics of a good e-textbook or e-material are:

- Online availability: the e-textbook/material should be available online, with the ability to download and use it without an internet connection.
- Flexibility: it must be flexible to meet the needs of individual teachers, students and groups of students.
- Cost-effectiveness: developing e-textbook is more expensive than developing a traditional textbook due to the additional features and technological requirements. However, considering the entire life cycle of the textbook, with all updates, corrections, and the possibility of using certain components in other e-textbooks and other teaching materials, the total cost should be comparable or even lower.
- Sustainability: the e-textbook/material must allow adaptation to technological changes.
- Interoperability: the e-textbook should be accessible in different learning environments and with different tools.
- Usefulness in different pedagogical situations: In addition to use in face-to-face teaching, laboratory exercises, group work, flipped learning, homework, it is important to use and adapt individual parts of several e-textbooks to create a customised version of the e-textbook.

The following recommendations are well considered when creating e-textbooks [7, 8, 13]:

- Content and format must be separate.
- The material that is part of the e-textbook must be modular, allowing content to be added and removed.
- It is desirable that the technologies used are based on open source code.
- E-textbooks and all their parts must be transferable to different learning environments (e.g., online classrooms, learning management systems).
- E-textbooks within a given school environment should have a simple and consistent user interface.

We can conclude that the e-textbooks and e-materials used by teachers and students learning and teaching CS in Slovenian schools, presented in this paper, take into account most of the features and recommendations mentioned above. To a lesser extent, these materials can be adapted to individual needs and context of use [13, 15].

CS teachers emphasize the importance of integrating interactive tasks into e-textbooks and the alignment of the content of textbooks with the subject curriculum. They believe that e-textbooks and interactive tutorials help them teach the subject matter, consolidate knowledge, check students' knowledge, as well as provide immediate feedback. Teachers miss high-quality materials for teaching and learning the elective CS subjects in the second educational cycle and materials for data and analysis, and CS effects on society, especially on popular topics such as artificial intelligence, security, and cryptography. Teachers would like to get examples of lesson plans for individual topics and more materials in Slovenian. Teachers would also like to see more examples and exercises of varying complexity.

In the future, e-textbooks and interactive materials in Slovenian should be provided, covering all topics from the curricula and regularly updated in line with the research in Computer Science education.

References

1. Anželj, G., Jerše, G., Lokar, M.: Blockly, pišek in poučevanje programiranja = blockly, pišek and teaching programming. In: Rajkovič, U., Batagelj, B. (eds.) Education in information society - VIVID 2018: Conference Proceedings of the 21st International Multiconference Information Society - IS 2018, pp. 11–19. Založba UL FRI (2018). http://zalozba.fri.uni-lj.si/VIVID2018.pdf
2. Bell, T., Vahrenhold, J.: CS unplugged—how is it used, and does it work? In: Böckenhauer, H.-J., Komm, D., Unger, W. (eds.) Adventures Between Lower Bounds and Higher Altitudes. LNCS, vol. 11011, pp. 497–521. Springer, Cham (2018). https://doi.org/10.1007/978-3-319-98355-4_29
3. Brodnik, A., et al.: Snovalci digitalne prihodnosti ali le uporabniki?: poročilo strokovne delovne skupine za analizo prisotnosti vsebin računalništva in informatike v programih osnovnih in srednjih šol ter za pripravo študije o možnih spremembah (RINOS). Ministrstvo za izobraževanje, znanost in šport, Ljubljana (2018)
4. Demšar, I., Demšar, J.: Računalništvo brez računalnika. In: Juriševič, M. (ed.) Motiviranje nadarjenih učencev za učenje naravoslovja: zbornik povzetkov. p. 38. Pedagoška fakulteta (2013)
5. Edgcomb, A., Vahid, F., Lysecky, R., Lysecky, S.: Getting students to earnestly do reading, studying, and homework in an introductory programming class. In: Proceedings of the 2017 ACM SIGCSE Technical Symposium on Computer Science Education, SIGCSE 2017, pp. 171–176. ACM, New York (2017). https://doi.org/10.1145/3017680.3017732
6. Edgcomb, A.D., Vahid, F.: Effectiveness of online textbooks vs. interactive web-native content. In: 2014 ASEE Annual Conference & Exposition, pp. 24.460.1-24.460.10. ASEE Conferences, Indianapolis, Indiana (2014). https://doi.org/10.18260/1-2-20351
7. Ericson, B.: An analysis of interactive feature use in two ebooks. In: Sosnovsky, S.A., Brusilovsky, P., Baraniuk, R.G., Agrawal, R., Lan, A.S. (eds.) Proceedings of the First Workshop on Intelligent Textbooks co-located with 20th International Conference on Artificial Intelligence in Education (AIED 2019), vol. 2384, pp. 4–17. CEUR-WS.org, Ulm (2019)
8. Ericson, B.J., Rogers, K., Parker, M., Morrison, B., Guzdial, M.: Identifying design principles for CS teacher ebooks through design-based research. In: Proceedings of the 2016 ACM Conference on International Computing Education Research, pp. 191–200. ACM, New York (2016). https://doi.org/10.1145/2960310.2960335
9. Fouh, E., et al.: Design and architecture of an interactive etextbook-the opendsa system. Sci. comput. Program. **88**, 22–40 (2014). https://doi.org/10.1016/j.scico.2013.11.040
10. Jerše, G., Koren Ošljak, K., Lokar, M.: Poučevanje programskih konceptov: spletna zbirka nalog s samodejnim preverjanjem = teaching basic programming concepts: Online handbook with automated verification. In: Rajkovič, U., Batagelj, B. (eds.) Education in Information Society - VIVID 2019: Conference Proceedings of 22nd International Multiconference Information Society - IS 2019, vol. J, pp. 106–111. Institut "Jožef Stefan" (2019)

11. Jerše, G., Lokar, M.: Providing better feedback for students solving programming tasks using project tomo. In: Krusche, S. (ed.) Software Engineering Workshops 2018: SE-WS 2018: combined proceedings of the Workshop of the German Software Engineering Conference 2018 (SE 2018), pp. 28–31. CEUR-WS (2018)
12. K-12 computer science framework. https://k12cs.org/. Accessed 12 May 2022
13. Korhonen, A., et al.: Requirements and design strategies for open source interactive computer science ebooks. In: Proceedings of the ITiCSE Working Group Reports Conference on Innovation and Technology in Computer Science Education-Working Group Reports, pp. 53–72. ACM, New York (2013). https://doi.org/10.1145/2543882.2543886
14. Kotnik, K., Lasič, N., Lokar, M., Vogrinčič, R., Zdovc, M.: Praktične izkušnje pri poučevanju programiranja v srednji šoli z uporabo storitve projekt tomo = practical experiences in teaching high school programming with the projekt tomo platform. In: Rajkovič, U., Batagelj, B. (eds.) Education in Information Society - VIVID 2018: Conference Proceedings of 21st International Multiconference Information Society - IS 2018, pp. 143–150. Založba UL FRI (2018)
15. Lokar, M.: The future of e-textbooks. Int. J. Technol. Math. Educ. **22**(3), 101–106 (2015)
16. Miller, B.N., Ranum, D.L.: Beyond pdf and epub: toward an interactive textbook. In: Proceedings of the 17th ACM Annual Conference on Innovation and Technology in Computer Science Education, pp. 150–155. ITiCSE 2012, ACM, New York (2012). https://doi.org/10.1145/2325296.2325335
17. Mori, N., Lokar, M.: A new interactive computer science textbook in Slovenia. In: Brodnik, A., Tort, F. (eds.) ISSEP 2016. LNCS, vol. 9973, pp. 167–178. Springer, Cham (2016). https://doi.org/10.1007/978-3-319-46747-4_14
18. Pesek, I., Zmazek, B., Mohorčič, G.: Od e-gradiv do i-učbenikov. In: Pesek, I., Zmazek, B., Milekšič, V. (eds.) Slovenski i-učbeniki, pp. 8–16. Zavod Republike Slovenije za šolstvo, Slovenski i-učbeniki (2014)
19. Pollari-Malmi, K., Guerra, J., Brusilovsky, P., Malmi, L., Sirkiä, T.: On the value of using an interactive electronic textbook in an introductory programming course. In: Proceedings of the 17th Koli Calling International Conference on Computing Education Research, pp. 168–172. ACM, New York (2017). https://doi.org/10.1145/3141880.3141890
20. Shaffer, C.A., Naps, T.L., Fouh, E.: Truly interactive textbooks for computer science education. In: Rößling, G. (ed.) Proceedings of the Sixth Program Visualization Workshop, pp. 97–103. Darmstadt, Germany (2011)
21. Zhang, Y., Kudva, S.: Ebooks vs. print books: readers' choices and preferences across contexts. In: Proceedings of the American Society for Information Science and Technology, vol. 50, no. 1, pp. 1–4 (2013). https://doi.org/10.1002/meet.14505001106

Author Index